GOVERNING URBAN SUSTAINABILITY

Governing Urban Sustainability

Comparing Cities in the USA and Germany

LISA PETTIBONE
Museum für Naturkunde Berlin, Germany
and Freie Universität Berlin, Germany

Routledge
Taylor & Francis Group
LONDON AND NEW YORK

First published 2015 by Ashgate Publishing

2 Park Square, Milton Park, Abingdon, Oxfordshire OX14 4RN
711 Third Avenue, New York, NY 10017

Routledge is an imprint of the Taylor & Francis Group, an informa business

First issued in paperback 2018

British Library Cataloguing in Publication Data
A catalogue record for this book is available from the British Library

The Library of Congress has cataloged the printed edition as follows:
Pettibone, Lisa, author.
Governing urban sustainability : comparing cities in the USA and Germany / by Lisa Pettibone.
pages cm. -- (Urban planning and environment)
Includes bibliographical references and index.
ISBN 978-1-4724-6316-6 (hardback)
1. Sustainable urban development--United States. 2. Sustainable urban development--Germany 3 City.planning--Environmental aspects--United States. 4. City planning--Environmental aspects--Germany. 5. Urban policy--Environmental aspects--United States. 6. Urban policy--Environmental aspects--Germany. I. Title.
HT243.U6P47 2016
307.1'2160943--dc23

2015030417

ISBN 978-1-4724-6316-6 (hbk)
ISBN 978-1-138-54735-3 (pbk)

Contents

List of Tables and Box

Tables

Box

List of Abbreviations and Acronyms

German institutions' own translations of their names are used.

APA	American Planning Association
BMBF	German Ministry for Education and Research (*Bundesministerium für Bildung und Forschung*)
BMVBS	Federal Ministry of Transport, Building and Urban Development (Bundesministerium für Verkehr, Bau und Stadtentwicklung)
B-Nax	Berlin Sustainability Index (*Berliner Nachhaltigkeitsindex*)
BPA	Bonneville Power Administration
BPS	Portland Bureau of Planning and Sustainability
BROG	German Federal Land-Use Law (*Bundesraumordnungsgesetz*)
BSU	Agency for Urban Development and Environment (*Behörde für Stadtentwicklung und Umwelt Hamburg*)
BUND	Friends of the Earth Germany (*Bund für Umwelt und Naturschutz Deutschland e.V.*)
CBD	Convention on Biological Diversity
CDU	Christian Democratic Union
CEMR	Council of European Municipalities and Regions
CLF	Coalition for a Livable Future
CNU	The Council for the New Urbanism
CSD	UN Commission on Sustainable Development
DBU	German Environmental Foundation (*Deutsche Bundesstiftung Umwelt*)
DLCD	Oregon Department of Land Conservation and Development
DOT	U.S. Department of Transportation
DPD	Seattle Department of Planning and Development
EDF	Environmental Defense Fund
EJ	environmental justice
EPA	U.S. Environmental Protection Agency
ESD	education for sustainable development
FDP	Free Democrat Party
FEST	Protestant Institute for Interdisciplinary Research (*Forschungsstätte der Evangelischen Studiengemeinschaft e.V.*)
GAL	Green Allied Left, Hamburg's Green party
GDP	Gross Domestic Product
HEINZ	Hamburg's Indicators of Sustainable Development (*Hamburger Entwicklungs-INdikatoren Zukunftsfähigkeit*)

HUD	U.S. Department of Housing and Urban Development
IASS	Institute for Advanced Sustainability Studies
ICLEI	International Council for Local Environmental Initiatives
IFEU	Institute for Energy and Environment Research (*Institut für Energie und Umweltforschung Heidelberg GmbH*)
JESSICA	Joint European Support for Sustainable Investment in City Areas
LA21	Local Agenda 21
LULUs	Locally Unwanted Land Uses
MUE-25	Managing Urban Europe 25
NABU	Nature and Biodiversity Conservation Union (*Naturschutzbund Deutschland e.V.*)
NLC	National League of Cities
NRDC	Natural Resources Defense Council
NYC EJA	New York City Environmental Justice Alliance
OLTPS	New York Office of Long-Term Planning and Sustainability
PBNE	Parliamentary Advisory Council on Sustainable Development (*Parlamentarischer Beirat für Nachhaltige Entwicklung*)
PCSD	U.S. President's Council on Sustainable Development
PoSI	Portland Sustainability Institute; now EcoDistricts
PSC	Partnership for Sustainable Communities
RNE	German Council for Sustainable Development (*Rat für Nachhaltige Entwicklung*)
SPD	Social Democratic Party
SRU	German Advisory Council on the Environment (*Sachverständigenrat für Umweltfragen*)
SSBX	Sustainable South Bronx
SSU	Berlin Senatsverwaltung für Stadtentwicklung und Umwelt
STEP	Urban Development Plan (*Stadtentwicklungsplan*)
SWMP	Solid Waste Management Plan
UfU	Independent Institute for Environmental Issues (*Unabhängigen Institut für Umweltfragen*)
UGB	Urban Growth Boundary
UN	United Nations
UNCED	United Nations Conference on Environment and Development
UNCHE	United Nations Conference on the Human Environment
UNFCCC	United Nations Framework Convention on Climate Change
USGBC	U.S. Green Building Council
WBGU	German Advisory Council on Global Change (*Wissenschaftlicher Beirat der Bundesregierung Globale Umweltänderungen*)
WCED	World Commission on Environment and Development
WPPSS	Washington Public Power Supply System
WWSCI	*WirtschaftsWoche* Sustainable City Index

Acknowledgments

This book is the product of several years of research, writing, and revision. I would like to thank series editors Don Miller and Nicole Gurran for providing me the opportunity to join Ashgate's Urban Planning and Environment series. Thanks also to Valerie Rose and Margaret Younger at Ashgate for their work in bringing my manuscript to print.

The research on which this book is based would not have been possible without the guidance of Prof. Dr. Miranda Schreurs and Prof. Dr. Johann Köppel. The German Environmental Foundation (DBU) gave me the time and resources to complete this ambitious research project, as well as vital practice discussing my work in German. Harald Mönch at the FFU and Dr. Hedda Schlegel-Starmann at DBU were key in orchestrating my field research in the United States and Germany. While in the United States, researchers at the Earth Institute at Columbia University, the Evans School at the University of Washington, and the Hatfield School of Government and the Institute for Sustainable Solutions at Portland State University supported my research stays in those three cities. I also want to thank Hans Diefenbacher and his colleagues at the Protestant Institute for Interdisciplinary Research (FEST) for facilitating my research stay in Heidelberg. I am thankful to my interview partners in both the United States and Germany, who graciously took the time to share their expertise and insights with me. I also got valuable feedback from my colleagues at the FFU, other DBU scholars, and countless other academics I met along the way—especially those I disagreed with.

Thanks also go to my friends and family who supported me on this journey, and to my writing group, who coached me during the intense writing phase—and with whom I enjoyed so much quiche, cake, and coffee. This book was influenced by my work at Citizens For Europe; the emotional support and fresh perspective of the CFE team were a gift throughout. I want to thank the developers of Scrivener for making the bulk of the writing process so delightful. My most special thanks to André Krüger, who motivated me to finish this book, and who doesn't let me work too hard.

PART I
Defining Urban Sustainability

Chapter 1
A New Framework for Urban Sustainability Governance

Decision-makers today are faced with a number of interconnected challenges: a complex ecological crisis that includes biodiversity loss, multiple forms of pollution, and climate change; an increasingly precarious global financial system that has led to two recent crises, increased inequality, and growing social unrest; and growing tension between the Global South and the Global North about how to address these issues. But these challenges have been increasingly linked to solutions called for by sustainability, a worldview that has emerged from outside traditional policymaking channels and reframes these problems in a new way (Lafferty, 2004). Sustainability uses a reflexive, deliberative, systems-based approach to advance solutions that work within the long-term and acknowledge ecological limits. To achieve this, sustainability calls for a massive societal transformation (for example, McCormick et al., 2013; WBGU, 2011).

Such a transformation entails dramatic restructuring of social, political, and technical systems and institutions in order to reduce wasteful resource use, promote equity and inclusion, all while maintaining a good quality of life and without collapsing economic or ecological systems (WBGU, 2011). This tall order is the focus of research on sustainability governance, which focuses on transforming socio-political structures and processes within the framework of a broader transition to sustainability (Lafferty, 2004; Meadowcroft, Farrell, and Spangenberg, 2005). Sustainability governance requires new forms of decision-making to reframe problems and work in new ways to resolve them (Adger and Jordan, 2009; Voß, Bauknecht, and Kemp, 2006).

Partly because of sustainability's "outside-in" character (Lafferty, 2004)—that is, the fact that it is an agenda that has emerged largely from outside the political system—it has seen slow implementation at the national and international levels. In the face of this, cities worldwide have taken the lead in transitioning to sustainability, particularly in high-consumption countries such as the United States and Germany. Scholars have pointed to the importance of cities, as home to much of the world's population, pollution, and economic activity, in climate and sustainability governance (McCormick et al., 2013; Owen, 2009). In addition, cities' relative flexibility and ability to experiment with new policies has made them able to more quickly progress in a transition to urban sustainability governance (Acuto, 2013; Gordon, 2013). Although sustainability requires a global transformation, cities have taken the lead in moving toward sustainability governance. However, findings from urban sustainability governance research have been sobering. Researchers have cited implementation concerns (Kern, Koll, and Schophaus, 2007) and political shortcomings (for example, Gibbs and Krueger, 2007; Holgersen, 2014) as problems. Beyond this, much sustainability governance literature, including urban sustainability governance, overrelies on instruments, which have seen uneven implementation at best. If governance is to play a role in sustainability transition, it is important to better understand what effective urban sustainability governance looks like.

This book seeks to explore how six U.S. and German cities are incorporating sustainability principles into political decision-making processes, that is, how they are achieving urban sustainability governance. I explore these questions by asking why cities actually do develop and use plans and indicators and how well their vision of sustainability matches principles found in the literature. By examining six cities with similar contexts (large cities in industrialized federal countries) that have experimented with both strategic plans and sustainability indicators in different ways in terms of development process, motivation, and actors involved, I aim to better understand their role in urban sustainability governance.

By studying these six cities, I hope to answer the following questions:

- *Why do cities develop strategic plans and sustainability indicators?*
- *How do these governance instruments help incorporate sustainability principles into political decision-making?*

The focus here is thus two-fold: to understand both actors' motivations in developing these instruments and to explore how well they connect to urban sustainability governance, as measured by sustainability principles. While trying to answer these questions, I found one type of actor group not discussed in the literature to be consistently important. In several cities, civil society groups—in the form of think tanks, volunteer initiatives, or nonprofits—were central actors in the development and use of plans and instruments to achieve sustainability. However, the work of such organizations has not yet received explicit attention in the academic literature. Therefore, I also explore the role of these groups, which I call sustainability-minded institutions, in urban sustainability governance in the six case cities. They are the focus of a third research question:

- *How do sustainability-minded groups use plans and indicators to promote urban sustainability governance?*

In answering these questions, I hope to better understand not only the potential role of plans and indicators in urban sustainability governance, but what functions they actually perform. Specifically, I focus on how different motivations affect their use in urban decision-making processes, as well as how well these tools are connected to sustainability itself. In addition, I highlight the special role played by sustainability-minded groups as an important subject for further study.

This study uses a qualitative comparative case study approach. I examine the three factors—strategic plans, sustainability indicators, and sustainability-minded groups—in six cities in the United States and Germany considered sustainability leaders: New York, Portland, and Seattle in the United States and Berlin, Hamburg, and Heidelberg in Germany. In studying the effects of these three factors, I hope to build a more robust understanding of sustainability governance better able to aid practitioners in taking effective action.

Why the United States and Germany

This study uses a multiple case study approach that includes cities in both the United States and Germany.[1] These countries serve as interesting cases for study for several reasons. First, both are wealthy, high-consuming countries with problems typical of the global North and of interest to researchers worldwide. Both countries are important in international debate related to sustainability and environmental problems. Second, and important to the study of urban sustainability, both are federal countries whose cities have a relatively high level of autonomy. Third, cities in both countries are seen as sustainability leaders in the literature and have sought to brand themselves as interested in sustainability. The case cities are often considered role models, meaning that actors engaged in incorporating sustainability into urban decision-making look to these cases for guidance. These factors make the United States and Germany, as well as the six cities studied here, especially important to the study of urban sustainability governance.

At the same time, the United States and Germany differ in several important ways, as discussed in Chapter 3. Most relevant to this study are urban planning policy and the national-level debate regarding sustainability. These differences, however, play a surprisingly limited role at the city level. As such, although cities will be compared by country to highlight national-level factors of importance in incorporating sustainability into decision-making, comparisons are also made along other axes to explore other key factors.

The Six Cities

In this book, I use a qualitative comparative case study approach to examine how six large cities in the United States and Germany have used strategic planning and sustainability indicators as tools of urban sustainability governance (see Yin, 2009). The cases were selected to maximize a diversity of approaches to sustainability and uses of these two governance tools to incorporate sustainability

1 Similarities between these two countries have made them the subject of a rich body of comparative literature (for example, Dolowitz and Medearis, 2009; Lafferty and Meadowcroft, 2001; Light, 1999; Ralston, 2012; Rose-Ackerman, 1995; Schreurs, 2003).

principles into political decision-making (see Pettibone, 2014a for more details on the methods used). In addition, during the empirical research, I became aware of the importance of civil society organizations I call sustainability-minded groups, which became an additional focus of this research and are discussed in Chapter 7.

Three cities in each country were selected as the maximum possible for a qualitative design. To select the case cities, I first consulted city sustainability rankings in the United States (Karlenzig et al., 2007; Svoboda et al., 2008; Thompson, 2009)[2] and Germany (Dovern, Rickels, and Quaas, 2012; Siemens, 2011a). In the second step, I selected highly ranked cities from these lists that reflected a range in terms of geography, governance, and approach to sustainability (see Table 1.1).

All six cities presented here have been considered leaders for their work on sustainability. Berlin's Local Agenda 21 is a model of inclusive planning; Hamburg was the 2011 European Green Capital; Heidelberg's plan and indicators have won the city numerous awards; New York's ambitious sustainability plan has drawn international attention; Portland has used comprehensive planning to manage growth; Seattle's volunteer-created sustainability indicator set has been subject to international study. Each city has attempted to use these instruments in some way, but led by different actor groups with different motivations.

Each of the six cities selected has been considered an exemplar in its use of plans or indicators for sustainability: all have used these tools in some way to promote urban sustainability governance. What differs is what actors are involved—politicians, administrators, or civil society organizations—and what vision for the city these actors seek to realize through plans and indicators.

The case cities vary in terms of size, population, economic prosperity, and the built environment. Table 1.2 highlights key characteristics of the six case cities. As the table shows, the cities vary on all of the characteristics seen here. The main differences—such as population density and housing stock—are not necessarily based on country, but more closely correlated to city size. Surprisingly, Berlin and Hamburg have a large number of residents in single-family homes, comparable to the more suburban cities of Portland and Seattle. These cities are thus broadly similar in terms of general urban features. The following sections introduce these six cities by providing a brief overview of important historical and political context and developments.

2 It should be noted that subsequent study of the case cities and city ranking methodologies made me skeptical of their value beyond advocating for a particular understanding of sustainability. These three U.S. city ranking projects may have come to similar conclusions, as all efforts had ended by 2010.

Table 1.1 Case city overview: Focus of analysis

	New York	Portland	Seattle	Berlin	Hamburg	Heidelberg
Plans studied	*PlaNYC* (2007)	*VisionPDX* (2008), *The Portland Plan* (2012)	*Toward a sustainable Seattle* (1994, 2004)	*Local Agenda 21* (2006), *Berlin 2030* (2014)	*Wachsende Stadt* (2002, 2003), *Wachsen mit Weitsicht* (2010)	*STEP 2010* (1997), *STEP 2015* (2006)
Indicator sets studied	PlaNYC performance measures	Portland Plan, Oregon Benchmarks	Sustainable Seattle indicators, Cascadia Scorecard	*BerlinStudie,* B-Nax, LA21 indicators	HEINZ	Sustainability test, Sustainability reports, FEST
Key actors	Sustainability Watch; Sustainable South Bronx and Sustainable Flatbush; NYC EJA, UPROSE	1000 Friends of Oregon, Coalition for a Livable Future	Bullitt Foundation, Sightline Institute, Sustainable Seattle	Berlin 21	*Zukunftsrat*	BUND Heidelberg, FEST, IFEU

* For the sake of simplicity, acronyms and abbreviations are used here. For more detailed information on these plans, indicator sets, and actor groups, please see the relevant empirical chapters.

Table 1.2 Case city statistics

	New York	Portland	Seattle	Berlin	Hamburg	Heidelberg
Size *(sq mi.)*	302.64	133.43	83.94	342.74	291.57	162.05
Population *(in 2010)*	8,175,136	583,778	608,660	3,300,000*	1,786,448	305,322
Growth *(% pop change since 2000)*	+2%	+9%	+7%	No change	+4%	+3%
Population density *(pop per sq mi.)*	27,425	4,377	7,251	9,628	6,125	1,882
Diversity *(% white, non-Hispanic (US); % German citizens (GER))*	33.3%	72.2%	66.3%	86.3%	86.9%	67% (does not include naturalized migrants)
Housing stock *(# housing units in 2010; % multi-unit)*	3,371,062 83.6%	265,429 38.1%	308,516 50.5%	1,894,564 49.7%	890,881 41.3%	72,307 (in 2011) 89.5%
Wealth† *(Median HH income (US); Average per cap discret. income (GER))*	$51,865	$51,238	$63,470	€16,608	€20,807	€21,555
City government	Strong mayor; city council	Commission-based: mayor + 4 commissioners	Mayor-council; 9 at-large council-members	Berlin Senate is mayor + 8 senators; state parliament	Hamburg Senate is mayor + senators; state parliament	mayor; city council
Key scales of government	city; state	city; county; metro region; state	city; county; state	city-state; neighborhood	city-state; neighborhood	city; region; state

U.S. sources: U.S. Census Bureau (2012). German sources: Siemens (2011); Stadt Heidelberg and GEWOS (2013); Statistisches Amt Berlin-Brandenburg (2010); Statistisches Amt für Hamburg und Schleswig-Holstein (2014); Statistisches Bundesamt (2011). *The table has Berlin's population at 3,460,725, but the city's census data released one year later was shown to have overestimated the city's population (*Berliner Zeitung*, 13 May 2013). †Here the statistics should not be considered internationally comparable, as they differ in terms of content (earned vs. discretionary income) and units (household vs. per capita). Instead, they serve to highlight the relevant differences within countries.

Berlin

Berlin is the capital of Germany and its most populous city. The city boasts a comprehensive public transit system, a network of waterways, and an older housing stock. The city has a thriving cultural scene that includes numerous grassroots activities related in some way to sustainability. Its biggest challenge, however, is the city's financial situation, which has been described as "disastrous" (Kern, Koll, and Schophaus, 2007; see also Krätke, 2004). After a brief infusion of cash in the 1990s to reconnect the city's split halves, renovate socialist housing in the East, and rebuild large physical gaps in the heart of the city created by the Wall (Bernt, 2012),[3] Berlin was left to fend for itself. At the end of 2010, the city was over €61 billion in debt (Amt für Statistik Berlin-Brandenburg, 2014). This may be one reason why the city government has focused its attention in the last decade on economic development, particularly by seeking to attract major corporations and wealthy foreign investors as a "city of talents" (Krätke, 2004).[4] The Senate's development plans have led to public outcry in many cases, most recently leading to a 2014 public referendum that rejected plans to develop the public space on Tempelhof field (Bartlick, 2014; see also Zwangsräumung verhindern, 2013).

Berlin is a city-state (*Stadtstaat*), giving it the political representation of a state in the federal *Bundesrat.* The city has 12 neighborhoods (*Bezirke*) of relatively equal population (Statistisches Landesamt Berlin, 2012). The city's 140-member parliament (*Abgeordnetenhaus*) elects the mayor; the executive consists of up to eight senators selected by the mayor to lead city agencies. This has led to some friction between the executive and legislative branches in recent years. Berlin also works closely with the adjacent state of Brandenburg, which forms the metropolitan region's rural and suburban fringe.

Berlin could be described by its "creative sustainability": grassroots projects that create neighborhood tool libraries to increase sharing, create hotel rooms out of old mobile homes, and transform the city's oldest gas station into an artistic filling station that tops up creative energy (see also Berlin 21, 2014). Such projects are possible due to the city's relatively cheap real estate and history of grassroots neighborhood engagement (LaFond, 1999). Many Berliners prefer lifestyles consistent with sustainability: sharing, using few resources, and developing innovative lifestyle models that reinforce values of solidarity and sufficiency

3 The post-1989 reconstruction of Potsdamer Platz, a dynamic central marketplace in the 1920s that, divided by the Wall, became a barren wasteland in the 1960s, is an excellent example of this. After reunification, one plan was to build a massive sustainability center in the space (LaFond, 1999). Financial realities led to the construction of a massive tourist destination, with the Sony Center a main attraction. A visit today should be compared to the area's appearance in two iconic films: first as part of a bustling metropolis in *Berlin: Sinfonie einer Großstadt* (1927) and then as a vast wasteland in *Wings of Desire* (1987).

4 This trend could be linked to the more general neoliberalization of major cities (Hackworth, 2007).

(Berlin 21, 2014). The city's thriving urban gardening scene presents one example of this (Sandilya and Danna, 2013, 56). In addition, numerous nonprofit and for-profit start-ups exist in areas as diverse as car-sharing, tool libraries, and eco-housing (LaFond, 1999), as well as sustainable fashion and IT services.

Hamburg

Hamburg is the second-most populous city in Germany. Much of Hamburg's development is related to its port and shipping industry: a shift to container shipping in recent decades has led to the development of heavy, environmentally damaging infrastructure in the area surrounding Hamburg and a general suburbanization of the city. Like Berlin, Hamburg is also a city-state, which gives it a voice in federal decision-making in the German *Bundesrat*. Its state government is led by the mayor, who leads the city's executive branch, the Senate.[5] Hamburg's parliament (*Bürgerschaft*) includes roughly 120 members and elects the mayor. In addition, seven neighborhoods include their own decision-making bodies and jurisdiction (FHH, 2011).

Like Berlin, policy in Hamburg in recent decades has focused on economic development that draws international capital, such as Klaus von Dohnanyi's "entrepreneurial city" (Schindler, 2011, 168–9). After a brief center left (SPD)–Green coalition in the 1990s, that initiated a diffuse array of sustainability and climate measures, sustainability has served primarily as an instrument of economic policy in city politics. This is echoed in the city's strategic visions, discussed in Chapter 5. Current mayor Olaf Scholz has focused on pragmatism and implementation over grand visions; sustainability is not a priority (*SPD Regierungsprogramm*, 2011).

Despite the apparent lack of environmental interest, Hamburg was named the European Green Capital in 2011 by the European Commission.[6] Hamburg won

---

5 In Berlin and Hamburg, the city executive is called the *Senat*, which for simplicity is translated here as "Senate." The legislative branch of both cities is referred to as its parliament.

6 The award highlights a leading city on environmental and sustainability issues and primarily serves as a marketing tool for foreign investment and tourism:

> Being a European Green Capital brings many benefits long after the designated year ends. Some of the city specific benefits of our previous winners are detailed below. A summary of these includes:
>
> - Increase in tourism;
> - Positive international media coverage worth millions of euro;
> - Increase in international profile, networking and new alliances;
> - New jobs – a Green Capital is more attractive to foreign investors;
> - More emphasis on environmental projects through sponsorship and grants;
> - Pride among citizens;
> - Momentum to continue improving environmental sustainability
>
> (European Green Capital website, 2014).

the award for its renewable energy sector and transportation projects and used the award to leverage funding for a series of public events on sustainability issues that the city's environment agency hopes will become part of a sustainability strategy (HH01). Much like the European Green Capital Award, however, Hamburg's definition of sustainability appears to be focused on improving the city's reputation outside the city rather than incorporating sustainability principles inside the city, as will be seen in subsequent chapters. At the same time, the city-created independent Future Council (*Zukunftsrat*) has become a major player in sustainability in Hamburg.

Heidelberg

The idyllic city of Heidelberg is home to a major university, formerly housed a major U.S. military base, and attracts tourists from around the world to the ruins of Heidelberg castle. Much of the city's recent urban planning has revolved around redevelopment of the military base and the Bahnstadt, a depressed area southwest of the train station. The city is well connected to neighboring agglomerations through the regional rail system; it also has a bus system, several bike lanes, and a pedestrian corridor in the tourist-friendly downtown. Housing has been a recent political concern, especially due to the relatively high commuting rates from outside the city due to high rents and a tight housing market.

Heidelberg is the only German city studied here that is not also its own state. Instead, policy in Heidelberg is linked to the city government, the Rhein-Neckar region, and the state of Baden-Württemberg. The former coordinates transportation and other regional policies. Baden-Württemberg has been involved in sustainability issues for nearly two decades, first commissioning a sustainability indicator report with three other German states in the 1990s that has been updated several times (Diefenbacher et al., 2009; see Chapter 7). Since the Green party became the lead coalition partner (with the SPD) in the state government in 2011, it has undertaken a strategy in which sustainability is no longer considered an additional group of projects or programs with a dedicated budget, but should form the basis for all political decision-making (HD02).

Some of the state-level strategy has come from Heidelberg's lead, as the city has been a forerunner on sustainability. Central to what the German Environmental Foundation (DBU) calls Heidelberg's "systematic" approach to sustainability is the city's urban development plan (*Stadtentwicklungsplan* or STEP), which it also considers its Local Agenda 21 (see Chapter 5). Heidelberg was also the first German city to develop a climate action plan, in 1992, and has won several awards at the national and European level for its activities related to sustainability and the environment. Beate Weber, the city's mayor from 1990 to 2006, won the prestigious German environmental prize in 2007 for her systematic support of sustainability in Heidelberg (DBU, 2007), primarily the city's STEP.

New York

New York City, as the country's most populous city, is an outlier in the U.S. context. Key features are the city's comprehensive transit network, high population density (nearly three times the next densest city studied here), and large tax base of major corporations (City of New York, 2007; Owen, 2009). At the same time, the city's physical infrastructure has led to air pollution, lack of greenspace, and the increased threat of catastrophic flooding, most recently by Hurricane Sandy in 2012. Decades without investment in the city's massive infrastructure have led to congestion and capital deterioration (City of New York, 2007). Neoliberal development policies similar to those seen in Berlin and Hamburg have improved the city's financial situation following the "white flight" of the 1970s that led to a financial crisis that caused a partial takeover of city government operations by New York State in 1975 (Berg, 2007).

The new century has seen the policy focus shift from stemming population decline to accommodating expected growth (for example, Aggarwala in Byles and Kazi, 2008, 12). Indeed, this is how sustainability found its way onto the city's political agenda: through economic development policy. The city's sustainability plan, PlaNYC, began as such, and seeks to reconcile development plans with climate policy (Bradley and Gallucci, 2013). The city's strong mayor system, as well as the political independence of its long-serving mayor Michael Bloomberg, allowed for the relatively smooth development of the plan and implementation of many of its initiatives. Beyond its borders, the city's conflicting priorities with more rural, conservative upstate residents have led to a tense relationship with the state government. The New York metropolitan area also includes New Jersey and Connecticut, which increases the complexity of regional land use and transportation planning.

Environmental issues are relatively less prominent in city policymaking, with the largest environmental interest group the environmental justice community. Groups such as the New York City Environmental Justice Alliance (NYC EJA) and UPROSE seek to reduce the inequitable distribution of environmental ills, such as construction of locally unwanted land uses (LULUs) primarily in low-income neighborhoods. During the Bloomberg administration, these groups were given increased policymaking power: Advocate Eddie Bautista gained a role in the administration (NY08) and several groups were asked to participate in the mayor's Sustainability Advisory Board, which helped develop the city's sustainability plan.

Portland

Portland has been the subject of numerous research efforts due primarily to its unique blend of civic engagement, strong urban planning policies, and interest in sustainability and environmental issues (for example, Ozawa, 2004; Portney, 2003; Putnam and Feldstein, 2003; Rosan, 2007). Portland is a small, suburban, relatively homogenous city. Although the city has 50 percent lower density than sprawling

Los Angeles (Richardson and Gordon, 2001), decades of land-use planning have preserved a thriving and relatively compact downtown, built miles of bike lanes, and created and preserved several mixed-use neighborhoods (Ozawa, 2004).

Much has been written about Portland's particular brand of environmentally minded and civically engaged population (for example, Ozawa, 2004; Putnam and Feldstein, 2003). This is due in part to local government: in addition to the six nonpartisan elected officials who form the core of the city government, political decision-making is closely linked to the Multnomah County and unique elected Metro government. Today city, county, and regional agencies have a high level of cooperation, dividing responsibilities to increase efficiency. Beyond this, historian Carl Abbott (2001, 2004, 2011) describes Portland's evolution from a wild place to an active communitarian enclave today. The turning point was the rise of grassroots community activity in the 1960s and 1970s, which focused on challenging the sort of car-centric development taking place throughout the country. In particular, a protest movement led by the group Riverfront for People successfully challenged plans to expand a highway along the city's western riverbank; instead, the existing highway was removed to make way for a public park. Putnam and Feldstein (2003) describe Portland's high civic engagement as "a positive epidemic." Johnson's (2004) evaluation of the city's engagement finds that although it was indeed high in the past decades, both active engagement and the civility of public discourse have declined since the 1980s. At any rate, the city's self-selected, fairly homogeneous, environmentally minded population presents narrower extremes than a more diverse city such as New York (Ozawa, 2004).

Sustainability in Portland is closely linked to its comprehensive growth management policies. Since the 1970s, the state of Oregon has had perhaps the strictest growth management legislation in the country (Seltzer, 2004), requiring that cities and counties create Urban Growth Boundaries (UGBs) to concentrate development and protect farmland from development. Portland's UGB was created in 1979 and is maintained by Metro, an elected regional government that is unique in the United States. The UGB has contributed to the revitalization of its downtown, as well as mitigated the sort of sprawl that Oregonians dread: their nightmare is to become the next asphalt-covered, car-dependent Los Angeles (PDX03). This has created an interesting coalition of rural farmers and city dwellers, who have historically seen the same policy as serving both their interests (see Chapter 7). In recent years, however, this original supporting coalition has fractured and a growing suburban population interested in sprawling development has increased (for example, Ashworth in Oates, 2006). Planning requirements were dramatically weakened in 2004 with the passage of Measure 37,[7] which was scaled back in 2007. Pressures on Portland's UGB will likely increase as available land within the boundary becomes scarcer, affecting support of the policy.

---

7 Measure 37 required state and local governments to compensate landowners for lost property values due to land use regulation, essentially exempting private landowners from planning laws and regulations (de Place and Kaleda, 2007; Portney, 2003, Loc. 1436–9).

In addition to its ambitious planning policies, Portland has been a leader on environmental sustainability and energy policies. The city first developed a set of Sustainable City Principles in 1994 and was the first city in the country to create a carbon-dioxide reduction plan, in 1993, which was updated in 2002 and 2009 (Rutland and Aylett, 2008). Rutland and Aylett (2008) link the Pacific Northwest's ambitious climate goals to an energy policy that began with hydroelectric dams built as part of the New Deal and began to prioritize energy efficiency in the 1970s.

Seattle

Seattle is a mid-sized city in the Pacific Northwest. The city has a long history of civic engagement, environmental activism, and politically strong neighborhoods. Seattle is proud of its reputation as a "livable" city, born out of the forced economic diversification after massive layoffs from Boeing in the 1960s (Brambilla and Longo, 1980). Since then, Seattle has placed a high value on long-term, holistic planning that preserves the region's natural beauty and restrains the rate of growth while improving its high quality of life, for example with the Growth Management Act passed in 1990. Densification has been an area of strong contention, as residents fear that increased density will reduce their prized quality of life. In many respects, the city is a near mirror image of New York: it is a city of environmentalists who live in a city built to minimize environmental benefits even as its residents clamor for sustainable public and private solutions.

Seattle is generally tied with Portland, Oregon, at the top of rankings of sustainable U.S. cities (see, for example, Karlenzig et al., 2007; Portney, 2003). The city began to use sustainability as a principle for its state-mandated comprehensive planning in the early 1990s, making it an early adopter of the term (see Chapter 5). In addition, Seattle's political history over the past several decades has created an engaged citizenry interested in collaboration to plan for the future and concerned with quality of life and the environment (Brambillo and Longo, 1980; Sanders, 2010). Many principles of sustainability, especially those at odds with some American public values such as communitarianism and an interest in nature for its own sake, have deep roots in Seattle (Sanders, 2010). The centrality of such values to the public leads to an easier acceptance of—or even demand for—political decisions explicitly tied to sustainability.

Seattle uses a council–mayor political system with nine at-large councilmembers elected on a nonpartisan basis. Its political style is decentralized and consensus-based, which long-time local columnist Berger has characterized as a strength and a weakness (Berger, 2009, 238–9). This approach stands in marked contrast to the political style in New York, but is again similar to Portland. Seattle's relationship to higher levels of government is generally less antagonistic than New York's, but friction with Republicans in the state legislature has recently led to problems with the city's public transportation budget (Goldy, 2013). Like Portland, Seattle cooperates with county and regional governance structures. The city is the seat of King County, with which it has collaborated on climate and sustainability policies (see Chapter 5). In addition, Puget Sound Regional Council, although not as powerful as Portland's Metro government, has helped coordinate regional transportation, land use, and economic development planning.

Summary: Sustainability in Six German and U.S. cities

The six cities that are the focus of this research have numerous differences in terms of size, urban form, political institutions, and more. However, these brief summaries highlight two key features that make these cases worthy of study. First, all six cities have used strategic plans and sustainability indicators to address governance problems. Second, and just as important, these cities have used these tools for different reasons and to achieve different ends. These two features make these cities interesting cases to explore in understanding the role of strategic plans and sustainability indicators in urban sustainability governance. In addition, the different actor constellations in each city present the additional opportunity to explore the role of sustainability-minded groups as well.

Research Approach and Design

As stated in the last section, this book uses a qualitative case study approach. As is typical in case study research, I relied on multiple types of data (Yin, 2009), primarily gray literature, previous academic research, and interviews. The foundation was the gray literature on sustainability in each city, including sustainability reports and indicators, strategic plans, background documentation, and publications from sustainability-minded groups. The existence, quality, and availability of such reports varied widely between cities, reflecting both how sustainability has been incorporated into city decision-making processes as well as the cities' interests and capacities. In addition to the gray literature, I used academic publications on the case cities, which also ranged widely, from the excellent work done by Delia Schindler (2011) in Hamburg and Meg Holden (2006, 2007, 2008) in Seattle to an almost total lack of published scientific research on Heidelberg, despite disbelief from several interview partners that this could be the case.

Finally, I conducted semi-structured interviews in each city with actors working on sustainability-related issues in the city (following Feldman, 1981). These interviews aimed to answer questions raised by the background material, develop hypotheses and storylines, and better understand the interests of important actors involved in sustainability at the city level. Interviews generally lasted from an hour to 90 minutes and were conducted with a diverse range of stakeholders in order to get a range of views, following Glesne (1999), Yin (2009), and methods discussed in Rubin and Rubin (2005). Interview partners were primarily administrative or political actors (particularly in city planning or sustainability departments; members of sustainability councils, city council, and Local Agenda 21 groups); sustainability advocates (such as staff and volunteers at think tanks, nonprofit, or community groups); and university professors (who provided general background and context, important contacts, and independent research on relevant issues). I also spoke with several people working on urban sustainability issues at the national level in both countries, as well as some regional or state actors linked to the case cities. In all, I conducted 75 interviews between the fall 2011 and spring 2014 (see Table 1.5).

Table 1.3 Interviews conducted

	Berlin	Hamburg	Heidelberg	New York	Portland	Seattle
Interview code	BER	HH	HD	NY	PDX	SEA
Data collection period	Dec 2011, Nov 2013, Feb 2014	May 2013	Jun 2013	Jan–Feb 2012	Apr–May 2012	Feb–Mar 2012
Politicians, Administrators	1	2	4	5	4	6
Advocates	4	3	3	3	5	6
Other experts	2	2	2	7	6	10
Total interviews	7	7	9	15	15	22

In general, actors involved in sustainability in the six case cities were happy to talk to me, although in some cases key interviews were granted after building connections with previous interview partners. In rare cases, but surprisingly often in Heidelberg, potential interview partners declined interview requests. Although some cited time constraints, the more frequent reason given was that the person was not an expert on sustainability and thus unable to say anything meaningful.[8]

Information in the empirical chapters derived from interview material is coded with a city abbreviation (see Table 1.3, above) and a number of the interview partners there. Interviews with regional-level actors are coded with the relevant city; national-level interviews as "US" for the United States or "DE" for Germany.

As is true of any research project, this study has some important limitations. In particular, three issues should be kept in mind. First, qualitative case study research has been criticized for using a small number of atypical cases, and thus lacking external validity (cf. McKeown, 2004). However, as McKeown (2004) points out, the nature of such research is to develop new theoretical relationships. This research should thus be understood as exploratory. I hope to better understand why some cities in high-consumption federal countries use strategic plans and

8 This reason was also given in the German academic community, where I found a striking difference in approach from U.S. universities. In each U.S. case city, I found several professors knowledgeable and willing to speak to me on issues of sustainability in their city—indeed, university urban planning departments often worked closely with city government. German professors were both more reluctant to meet and less involved in local sustainability issues.

sustainability indicators, how these instruments are linked to sustainability principles, and how sustainability-minded groups contribute to urban sustainability governance. This is consistent with the purpose of case study research: to fully understand the individual cases studied (Creswell, 1998; Ragin, 2004; Stake, 1995). For this reason as well, I chose a large number of cases and multiple factors, which gives this study a "broad" rather than "deep" approach. Although this allows more robust findings in terms of what role different factors played in different contexts, it precluded me from examining any particular city in great detail. The exploratory, inductive nature of the research design instead focused on highlighting general trends across cities and areas of interest for future research.

A second limitation is the term "sustainability" itself. The restriction to actors and cities self-selected as interested in sustainability may have omitted important work by those who do not use the term itself (as found in Berke and Conroy, 2000). For example, Germany has a strong education for sustainable development (ESD) movement, which strongly links education policy to sustainability at all levels of government (DeHaan, 2006). Because the United States does not have the same interest in ESD, education was not as strongly related to sustainability in the U.S. case cities. The focus on self-identified sustainability advocates however served to better understand how sustainability itself is understood and advocated in the urban context. Future research, particularly that focused on the values or principles of sustainability, could include nested cases on the definition of sustainability within larger analysis of urban political discourse in general. For example, it could examine how policy problems are framed and the extent to which such frames are linked to sustainability principles. An additional area of analysis could be to determine the role played by the term itself in such framing. How does the term "sustainability" impact the incorporation of sustainability principles into public debate?

Book Content and Structure

This book examines the role of two governance instruments—strategic plans and sustainability indicators—as well as an actor group I call sustainability-minded groups in six U.S. and German cities. It is structured to explore the role of each in urban sustainability governance, rather than by case.

The rest of Part I sets the stage for the study. Chapter 2 defines sustainability as a contested concept and operationalizes it in this study as based on five core principles—acknowledgment of limits, interest in long time frames, systems approach, deliberation, and reflexivity—and discusses their relevance to the urban context. In Chapter 3, I discuss the historical context of the study, providing an overview of sustainability at the international level as well as in the United States and Germany. Finally, Chapter 4 discusses the concept of urban sustainability governance and how strategic plans and sustainability indicators are used to realize sustainability governance in cities, setting up the analytic framework used in the empirical research.

Part II presents the empirical findings on three discrete aspects of urban sustainability governance in six U.S. and German cities. Chapter 5 finds that plan implementation requires motivated actors to be successful. Although cities develop plans for multiple reasons, dealing with or encouraging growth is often a more central concern than sustainability. Chapter 6 points out different uses of indicators based on actor groups, and highlights the importance of developing indicator sets with such uses in mind. Finally, in Chapter 7, I discuss how sustainability-minded groups advance sustainability in the city context.

Part III discusses key findings from the cases and discusses their implications for urban sustainability governance research and practice. Chapter 8 analyzes the empirical findings and relates them back to the research questions: *Why do cities develop strategic plans and sustainability indicators, how do these tools allow them to incorporate sustainability principles into political decision-making, and how do sustainability-minded groups contribute to this process?* Chapter 9 concludes by summarizing the main points made throughout, discussing the implications of this research, and suggesting areas for further study.

Chapter 2
Sustainability in Five Principles

Sustainability is a complex concept that must first be understood before its implementation into political decision-making can be evaluated. This chapter defines sustainability such that its implementation in urban governance processes can be evaluated. In this study, I understand sustainability as an essentially contested concept based on five principles from the English- and German-language literature. Because sustainability is a contested concept, the importance of these principles—and at times even their definition—is debated by actors using the term. This means that sustainability cannot be evaluated as a simple, universally understood construct, but is constantly redefined within specific contexts.

However, simply saying that sustainability is contested does not allow for common evaluation across cities that may define it differently. For this reason, I outline five principles found in the literature as dimensions of consensus and debate. Specifically, I argue that sustainability represents a model of development that values long-term quality of life and uses a complex, systems approach to remain within ecological limits. Sustainability respects the values and specific problems of local stakeholders and prefers deliberative mechanisms to reach consensus on common goals. It also works as a reflexive, iterative process attuned to changes in circumstances and levels of uncertainty inherent to our incomplete understanding of the world. Further, I discuss how each principle of sustainability could be realized in the city context, aiding analysis of the case studies in Part II.

Understanding Sustainability as a Contested Concept

Many concepts critical to academic study contain properties that make them subject to continuous debate—in other words, contested. I follow Connelly (2007), who framed sustainability as consistent with what Gallie (1956) calls an "essentially contested concept." In brief, considering sustainability as an essentially contested concept allows it to be understood as a term constructed within a particular discourse, but also sets boundaries around its definition along lines set by that discourse.

Defining sustainability has become an academic exercise in itself. As the literature on sustainability has exploded, so has the number of books and articles that attempt simply to define the concept, or, as Conrad (2000) vividly describes it, "nail pudding to the wall." Because sustainability serves as an umbrella term for an idealized vision of the future, two large literatures have emerged that seek to nail down the term. First, a normative definitional literature argues for a particular meaning of or presents a framework for sustainability (for example, BUND and Misereor, 1996;

Coenen and Grunwald, 2003; Daly and Cobb, 1989; Farley and Smith, 2013; Ott and Döring, 2008). Second, a review literature seeks to define the term by presenting its history and key publications (for example, Brand and Fürst, 2002; Connelly, 2007; Conrad, 2000; Dresner, 2008; Kidd, 1992; Stockdale, 1989). Beyond attempts to define sustainability, some writers have sought to highlight particular aspects of sustainability (for example, Daly and Cobb, 1989; Jackson, 2009; Patel, 2009) or the importance of a transition to sustainability (AtKisson, 2009; Ehrenfeld, 2008; Luks, 2010). The volume of writing on the subject signals that sustainability is itself being contested, which Connelly (2007) acknowledges in describing sustainability as an essentially contested concept.

The term comes from Gallie (1956), and is worthy of a brief discussion. Gallie describes the "essentially contested concept" as one without solid definition. At its core, it is "not resolvable by argument of any kind, [but is] nevertheless sustained by perfectly respectable arguments and evidence" (169). The contention itself is a key characteristic of the concept: "there are concepts which are essentially contested, concepts the proper use of which inevitably involves endless disputes about their proper uses on the part of their users" (169). Gallie considers democracy, art, social justice, and religion classic examples of essentially contested concepts: everyone has an opinion on what these terms mean, but these opinions differ widely and cannot be reconciled into a single universal definition.

Gallie (1956) provides five characteristics of essentially contested concepts: 1) They signify a valued achievement 2) that is complex, in that it is based on several distinct characteristics. 3) The basic characteristics that together form the concept's value are generally agreed upon, but their contribution to the whole is the subject of debate. In addition, 4) the achievement itself is temporary, as the concept is constantly redefined to fit changing and unforeseen circumstances. Finally, 5) not only do different actors understand the concept differently, but they recognize the fact that it is contested by others and have some appreciation for other possible criteria. Gallie emphasizes the subjective aspect implicit in their use:

> to use an essentially contested concept means to use it against other uses and to recognize that one's own use of it has to be maintained against these other uses. Still more simply, to use an essentially contested concept means to use it both aggressively and defensively (172).[1]

1 Later in his essay, Gallie adds two more characteristics that will not be used here: 6) "derivation … from an original exemplar whose authority is acknowledged by all the contestant users of the concept" (in the case of sustainability, one could argue for the Brundtland Report or Limits to growth as the original exemplar) and 7) "the probability or plausibility … of the claim that the continuous competition for acknowledgement as between the contestant users of the concept, enables the original exemplar's achievement to be sustained and/or developed in optimum fashion" (Gallie, 1956, 180).

This has several implications. First, Gallie admits that the essentially contested concept precludes the existence of some general principle on which a winning definition can unanimously be determined. Instead, it acknowledges the rationality of different uses, individuals' changes in definitions, and users' own recognition of other definitions within a general framework (189–90).[2] Understanding sustainability as an essentially contested concept thus forces one to acknowledge a range of definitions based on how commonly agreed upon characteristics—such as environmental quality or social equity—are defined and weighted.

This suggests that, although an essentially contested concept may be judged through a number of vague and complex criteria, there are ultimately rules that govern its definition. The debate around sustainability exhibits several of the characteristics discussed by Gallie. First, sustainability itself is certainly a normative ideal that could be considered a "valued achievement." This achievement is indeed complex: the most common lay definition understands sustainability as the intersection between economic, environmental, and social concerns. The role each of these components plays is the subject of intense debate, but to some extent proponents of various perspectives respect the views of those they are debating. Finally, sustainability is considered by some to be a constantly changing target, both in the face of new scientific findings as well as changing values and political priorities (for example, BUND and Misereor, 1996; Brand and Fürst, 2002).

For this reason, disagreements on what is sustainable or not, and what sustainability itself is, must continue—they are part of the nature of the contested concept. At the same time, it should be possible to lay out common features of sustainability: those characteristics commonly agreed to play some role in its definition, even if their importance is subject to debate.

Here, I present five core principles of sustainability in order to demonstrate the concept's coherence and provide a foundation for testing their incorporation into political decision-making in U.S. and German cities. As such, this discussion seeks to be more broad than deep, highlighting the key arguments and trends in the English- and German-language literature on sustainability in the last 40 years. Each of the five principles is discussed in turn, along with illustrative examples of how each can be realized in urban sustainability governance.

The review presented here conflates the English- and German-language literatures to some extent. Although there are important differences of emphasis—for example, German authors have spent more time on notions of risk and equity, whereas more of the English literature looks for "win–wins" in sustainability—I would argue that both literatures feature the same principles. The differences lie primarily in the debates taking place within these principles and the weight they are given in advocating for sustainability. In other words, the differences are more of emphasis than definition.

---

2 Much of Gallie's discussion prefigures Habermas's (1985) theory of communicative action by calling for the expansion of reason based on exchange of ideas and implicitly valuing social, personal, and empirical qualities throughout his discussion.

Box 2.1 Sustainability in five principles

Based on the literature presented in this chapter, this study defines sustainability as including five core principles:

Acknowledges limits: The planet has overshot firm ecological limits.
Cares about the long-term: It is important to consider the well-being of humans and the planet long into the future.
Thinks in systems: The world is complex, interconnected, and impossible to fully understand.
Deliberative: All relevant actors must be engaged to develop a vision for the future and a plan to get there.
Reflexive: Although quick action is required, learning is critical and entails constantly challenging assumptions, checking progress, and changing course.

Natural Limits to Growth and Environmental Primacy

The first principle provides the foundation of sustainability: the existence of ecological limits. This concept runs counter to a key premise of classical economic thinking, which sees natural resources as substitutable and priced by the market in a way to protect them from overuse. It also forms the core of sustainability's urgent call for transformation. Taken essentially as a given within the sustainability literature, this principle is perhaps the most controversial outside advocates' circles.

The notion of limits has been consistently reasserted throughout the history of Western civilization, but advocates of limits have generally lost ideological debates to those who preach abundance (for example, Myers and Simon, 1994). Thomas Malthus sounded an early warning about human overpopulation, worrying that population growth would lead to pressures on the food supply, economic stress, and a reduction of living standards (Dresner, 2008; Grober, 2010). Early economist John Stuart Mill also worried about overpopulation and called for a "stationary economy" (Dresner, 2008). Beginning in the 1960s, such fears re-asserted themselves in publications such as *The Population Bomb* and *Silent Spring* (Dresner, 2008). The environmental movement during this period led to an increased awareness of and concern with the negative consequences of technological progress.

Perhaps the most famous and influential publication around this time was *The Limits to Growth*. In the 1970s, a group of systems thinkers were called upon by the Club of Rome to write about the growing problems caused by industrialization: increasing air and water pollution, systemic poverty, and exhaustion of natural resources. Meadows et al. (1972) developed a computer model that invariably predicted social and economic collapse if population and resource use continued to grow (see also Meadows et al., 2004). Their model showed that even technological advances could not compensate for the loss of ecological services, as increased reliance on technological fixes would eventually be exhausted by resource limitations, whether in terms of material or human capacity. In their report, the authors discussed the natural "limits to growth" and called for political change to move toward a more sustainable path. Daly (1977), working from the same premise but using the laws of physics, discussed the implications of entropy in the global economy: natural resources must be understood as scarce and thus used thoughtfully. He called for a "steady-state economy" that transitioned to renewable resources and kept their use at replenishment levels.

This concern stands in opposition to the ideology of abundance, espoused especially by neoclassical economists (for example, Cass, 1965; Solow, 1956), which argues that scarce natural resources and ecosystem functions can simply be substituted by human capital or technological innovation (Daly, 1987). Current economic thinking works from models in which free markets allow for the constant equilibrium of supply and demand, as well as easy substitution of scarce goods.[3] Myers and Simon (1994), in a public debate at Columbia University, highlighted the tension between this economics-led view and sustainability—framed in the event's title as *Scarcity or Abundance?* In the debate, Simon (an economist) would not accept Myers's evidence of the current extinction crisis and kept returning to figures of lowered resource prices as evidence of abundance. Myers, a biologist, could not successfully rebut Simon's economics statistics, and found the uncertainty inherent in his species data unconvincing to his sparring partner. The debate highlighted the revolutionary nature of this core principle of sustainability. Not only does it go against mainstream economic thought, which serves as the foundation for much of scientific and political thinking, but it suggests a major crisis that deserves action. Implicit in acceptance of limits is an acknowledgment that we have overshot those limits. This has been expressed most recently in Rockström's work on planetary boundaries. He and others recently argued that the planet has overshot these boundaries in three of its nine biological systems: biodiversity loss, climate change, and the nitrogen cycle (Steffen et al., 2015).

Accepting natural limits upsets the status quo, because it gives urgency to environmental concerns. Kopfmüller et al. (2001) highlight the differences between models of sustainability based on whether environment protection is the key value to protect or only considered in tandem with economic and social concerns. Farley and Smith (2013) advocate for a one-dimensional sustainability in their definition of "neo-sustainability": they use a nested model in which the economic and social spheres reside within the all-encompassing environmental sphere. These models are more advanced representations of the sustainability triangle—which seeks to balance environmental, social, and economic factors—ubiquitous in PowerPoint presentations on sustainability. Meadows (1998) uses the same metaphor as Farley and Smith in a slightly different model, a pyramid that moves from "ultimate means"—the natural environment—to the "ultimate ends" of self-actualization, through the middle stages of built, human, and social capital (42). These authors all place limits as the foremost concern of sustainability.

---

3 This premise has been criticized by sustainability thinkers for decades (for example, Daly and Cobb, 1989; Welzer and Wiegandt, 2014).

However, the notion of limits is difficult to operationalize in a policy context. This is especially apparent in open systems such as cities. Rees (1997) argues that urban sustainability may be an "oxymoron," as urban policy rarely takes into account all of a city's effects on the biosphere, for example the agricultural land required to grow its food. Current practices only shift environmental destruction away from the city, obscuring ecological ills from those responsible for them. Rees's sustainable urban policy regime calls for cities to increase their own self-reliance and be reconceptualized as "whole systems" that include the neighboring rural areas that produce the goods and services used inside the city. In addition, urban metabolism presents a different approach to the notion of limits to the city level by providing a more comprehensive understanding of resource use (for example, Pincetl, 2012; Zhang, 2013). This research focuses on flows of wastes and resources such as water, materials, and food to provide a better understanding of cities' net effects on ecological systems.

These approaches suggest two ways in which cities can reflect the principle of limits. First, attempts to reduce total resource use (especially on a per capita basis, to account for population changes)—whether in the form of energy, waste generation, or consumption—are consistent with the principle of limits. Such policies are different from those based solely on increased efficiency or substitution of one type of resource for another, which only kick the sustainability problem down the line and reflect a worldview of limitless supply. Second, cities can better account for limits by connecting their resource use to broader systems, for example by linking water use to regional watershed capacity or food consumption to nearby agricultural land. Efforts that seek to close loops, such as regional agriculture programs or watershed protection, would help keep cities' consumption of such resources within the limits of local availability. Application of this principle to cities therefore requires an understanding of the city's effects in the natural world both within and beyond its borders.

Extending the Limits: Down with Economic Growth?

One aspect of the limits principle is worth additional mention, as it is both particularly controversial within the literature and has been present in political debates in the United States and Germany.[4] This is the notion of economic growth. Growth as a goal has been contested at least since the works of neoclassical economists Galbraith (1958) and Mishan (1967). It is even more controversial among sustainability advocates, perhaps partly because of its direct attack on major economic policy goals in the industrialized world. Debates have increased in recent

4 The controversy can be seen in the 2013 final report of the German Enquete Commission on growth and well-being, where debate became so intractable that pro- and anti-growth camps published separate reports (A. Hassel at public event, 15 Apr 2013).

years and range from how to develop a new indicator to replace GDP[5] (Cobb, Halstead, and Rowe, 1995; Diefenbacher and Zieschank, 2011) to whether growth is an active problem or should simply be ignored (Jackson, 2009; Latouche, 2010; debate between Kallis, 2011 and van den Bergh, 2011). In Europe, increased efforts have been seen in recent years to move to a "post-growth" society, with a larger focus on how to transition and the effects of doing so (for example, Welzer and Wiegandt, 2014).

Heinberg (2011) uses the earth's biological limits to argue that the global economy has reached the end of its growth period and will now remain stable or contract. For him, the economy follows the same rules as other natural systems, but these rules have been ignored by economists who use substitution and efficiency to argue that infinite growth is possible (for example, Cass, 1965; Solow, 1956). Jackson's 2009 report, *Prosperity without Growth*, is a penetrating look at the implications of a move away from that includes policy proposals. Latouche (2010) envisions a future society in a world without growth as one that achieves reduced consumption with expanded public services, a sharing economy that replaces individual ownership of things like cars and tools, and heavy taxes on financial transactions and advertising, which to him promote harmful consumption and waste. All of these authors focus on building new visions in a society no longer seeks economic growth.

In short, economic growth is linked to forms of consumption that exceed planetary boundaries; to keep within ecological limits, economic growth must be removed as a political goal or even stopped outright. Not all sustainability advocates are critical of economic growth, which is why it has been the source of so much debate (for example, between proponents of "strong" vs. "weak" sustainability, although O'Neill, 2009, sees this debate as artificially inflated). At the same time, the fact that growth is being challenged at all is worth noting, as economic growth is a central political goal in virtually all countries and many cities. Much political decision-making in the U.S. and Germany revolves around this goal, which is evident in argumentation around policies' effects on jobs, economic competitiveness, and GDP. News media in both countries conflate economic well-being with stock market growth or a rise in average income, which has caused increasing consternation with a public that does not necessarily feel better off when these indicators increase (Cobb, Halstead, and Rowe, 1995).

The desirability of economic growth and its relevance to sustainability are particularly important at the city level, where growth policies have been a key feature of cities' competition for population, jobs, and investment (Hackworth, 2007). Molotch (1976) calls this the "urban growth machine," in which a city must continue growing in order to satisfy powerful place-based political constituencies such as real estate investors. Such competition often hurts losing cities, and has been called unsustainable or a zero-sum game (Bauriedl, Schindler, and Winkler,

---

5 Gross Domestic Product is an economic indicator intended to mention a nation's production of goods and services, but more often politically used as a proxy for economic well-being.

2008; Granholm and Mulhern, 2011). Hackworth (2007) connects urban growth policy to neoliberalism, linking it to the increased internationalization of capital and policies based on increasing a city's appeal at the expense of its neighbors (Hahn, 2010; cf. Florida, 2004). Hackworth does not mention sustainability as a form of resistance to the neoliberal city, but this may be because sustainability and growth are often considered compatible by local political actors (for example, in Hamburg and New York in this study). Others, such as Owen (2009), have called for increased urbanization, seeing cities as more resource-efficient—and thus sustainable—than suburban or rural areas. To him, urban growth at the expense of inefficient suburbs is compatible with sustainability.

This debate echoes concerns mentioned in the previous section: how to deal with open systems like cities in a framework that requires evaluation of the entire system. City policies that reflected a sustainability perspective would likely not measure progress by economic growth, but with indicators such as subjective well-being or affordability. In addition, cities moving away from an economic growth paradigm would rely more on resources within the existing system—that is, their current population and financial capital—rather seek to import more from outside the city, as policies dedicated to the latter would serve only to improve one city's standing at the expense of other areas.

Long time Scales

The imperative to heed ecological limits gives sustainability another characteristic that distinguishes it from mainstream political debate: the consideration of long time scales. One important discussion in the sustainability literature is the appropriate length of time by which to consider a system "sustainable." Suggestions range from several decades to "perhaps into the indefinite future" (Meadowcroft, 1997, 169; see also Bossel, 2001; Daly, 1977). Indeed, one German translation of sustainability is *Zukunftsfähigkeit*—the ability to exist into the future (see, for example, BUND and Misereor, 1996; also, Brand and Fürst, 2002, for a discussion of German terms). Logically, the acceptance of ecological limits would not require action without an interest in the future. A more extended timeframe is thus a core principle of sustainability.

Like the first principle, however, the consideration of long time scales is also controversial beyond sustainability advocates. The controversy surrounding this principle can best be seen in attempts to evaluate long-term consequences of policy, which has traditionally been done using economic-based analyses that use a discount rate. Daly and Cobb (1989) discuss the limitations of this approach, highlighting particularly the implicit normative assumptions behind the discount rate, which is often pegged to market interest rates and essentially devalues the future in relation to the present. In essence, discount rates are an economic tool used to understand how much something in the near or distant future would be worth today. In practice, analyses that use this method create a bias towards resource exploitation, as the increase in short-term wealth outweighs the long-term impacts on the environment.[6] The principle of considering long time scales would thus be consistent with low or even negative discount rates, and stands in marked contrast to the norms of current economic analysis (Daly and Cobb, 1989).

In terms of urban policy, sustainability's extended temporal outlook makes it compatible with urban planning, whose focus on the built environment and subsequent use of comprehensive urban development plans means longer time scales than used in most other policy sectors (Feitelson, 2004). As urban infrastructure such as buildings, water systems, and transportation can see productive use for a century or more, it is generally considered in a longer-term perspective than national-level policy areas such as defense, health, or even environmental policy. Here, cities with a strong planning culture may be ahead of national governments in their incorporation of this principle. In addition to urban planning, this principle has often been linked at the policy level to sustainable finance: cities should not run budget deficits, and should make budgetary decisions that generally invest in the city's infrastructure (Grabow et al., 2011). A city that incorporates long time scales into decision-making would thus weigh short-term effects against possible outcomes in the long term (at least 20 years). In addition, decision-makers could take steps to institutionalize long-term policy efforts, inoculating them from the partisan winds, by mandating such considerations in planning and other decisions. In short, a sustainable city will plan for the long term, not at its expense.

Systems-thinking

Whereas the first principles help clarify the goals of sustainability, the third focuses on the style of thinking required. Systems thinking is a heuristic that provides structure to sustainability advocates' worldview. Like sustainability, it shares roots in the field of ecology. It seeks to connect humans to the natural world and can be skeptical of the technological progress promoted by modernism (Bowers, 2000; Ehrenfeld, 2008). Systems thinking has been advanced by numerous sustainability advocates (for example, Espinosa and Walker, 2011; Innes and Booher, 2010; Meadows, 2008).

The Cartesian or modernist worldview, which arose during the Enlightenment, includes several important features (Bowers, 2000; Dresner, 2008; Grober, 2010; Wheeler, 2004). Humans are seen as separate from nature and should seek to dominate it. The world is a machine—it can be broken down into smaller parts that are studied individually and can be changed without affecting the larger system. Science and technology can help us do this by giving us necessary information and

6 Dietz and Neumayer (2009) link the controversy surrounding two major reports on climate change to discount rates. They argue that the divergent conclusions in reports by Bjørn Lomberg and Nicholas Stern stem from different discount rates used: 3–5 percent vs. 1.5 percent, that is, the extent to which these different authors discounted the future.

allowing us to progress, by increasing output and efficiency. Thus, neutral experts are required who can dissect problems and propose ways to solve them.

Wheeler (2004) argues that sustainability requires a break from modernism. He sees sustainability as the latest iteration of a long-standing resistance to a positivist, universal model seeking to maximize growth through technology and efficiency. Others, such as Dresner (2008), Ehrenfeld (2008), Bowers (2000), and Voß and Kemp (2006), echo this argument. Voß and Kemp (2006) raise concerns that the modernist paradigm leads to a style of problem-solving riddled with externalities: the unintended negative consequences of rational policy choices (5). Ehrenfeld (2008) distinguishes sustainability as seeking to "resolve" problems at the systemic level rather than "solving" individual parts. Indeed, recent writings on the sustainability transition discuss the need to change social-ecological systems, emphasizing a systematic rather than programmatic approach (for example, McCormick et al., 2013; Voß, Bauknecht, and Kemp, 2006; WBGU, 2011).

The systems thinker, in contrast to the modernist, sees the world as composed of interconnected, complex systems that are impossible to fully understand. The world is dynamic: systems interact in unforeseen ways that cannot be understood in pieces. Instead, a holistic approach is needed to observe interactions and unintended effects of changes. Complexity leads to a sense of humility: we should not presume that our interventions in complex systems will have simple linear outcomes, nor that the systems themselves will behave in obvious ways. In addition, systems thinking places humans within natural and social systems, implicating us in environmental changes such as climate change and biodiversity loss. They can thus be linked to the precautionary principle, which urges caution in the face of unforeseen effects (Stirling, 2009).

Complexity theory and systems thinking lead to several key features relevant to sustainability governance, including at the city level. First, a systems view requires a governance approach that is integrated and coordinated: rather than creating different agencies to deal with discrete policy problems, an overarching strategy is required to coordinate efforts and look out for unintended effects in other policy sectors. In addition, this approach is more skeptical about technology than the modernist view. Although there is discussion over the role of technology in a sustainability transition, most consider that technology must serve a larger purpose and see many technological advances as accelerating environmental destruction and other negative effects (for example, Owen, 2011). Finally, acknowledgment of complexity calls for humility, thus increasing the value of more modest policy approaches such as the precautionary principle.

The principle of systems thinking has several implications for urban sustainability governance. First, it highlights the importance of a general political strategy. This can be realized through instruments such as a strategic plan (see Chapters 3 and 5) or institutional arrangements such as a special office to coordinate strategy across agencies. The policy approach should also seek systemic rather than technological change: a sustainable transport strategy would not simply replace gasoline-powered cars with electric vehicles, but change how people moved around the city and the diverse effects of mobility patterns (such as on air quality, levels of fitness, and social equity). A governance approach incorporating this principle would include diverse policy sectors and stakeholders in the policy development process in order to be more fully aware of all of the policy's effects on the system. Finally, it would consider risks inherent in systemic change and use a more precautionary approach to policy.

Deliberative

The first three principles conceptualize sustainability as a worldview that seeks to preserve the integrity of social-ecological systems over long periods of time in the face of uncertainty and complexity. The fourth principle of sustainability—deliberation—connects to the these principles in several ways. First, it acknowledges the concept's inherent normativity: sustainability describes the future that we want, which must be defined by us. Deliberation is also a methodological requirement for practical reasons. The need for massive systemic change means that a sustainability transition cannot be steered by a tiny elite, but must motivate the entire population (Pettibone, 2014b).

For these reasons, sustainability has shown itself to be receptive to emerging notions of participation and deliberation in social theory, such as Habermas's communicative rationality, the constructivist paradigm, and citizenship theory (for example, in Dobson, 2009; Innes and Booher, 2010). At the same time, this principle raises several questions and has led to different opinions on the proper form of participation or deliberation and divergent approaches to engagement (Meadowcroft, 2004). For example, should the general public be consulted, or a targeted group of stakeholders? A truly deliberative process must have an open outcome: can sustainability thus be assured in such situations? Still, deliberation functions especially well at the local level, where a smaller number of actors work in close physical proximity and likely share many common values (Innes and Booher, 2010).

Deliberation has been a key aspect of sustainability in political documents since the beginning. Participation is one of the four pillars of sustainability as conceptualized by Agenda 21; one of the report's four sections addresses how to include major stakeholder groups (UNCED, 1992). Local Agenda 21, an important early framework for urban sustainability based on Agenda 21, explicitly includes the need for participatory decision-making (UNCED, 1992). Here, participation is seen as a way to build ownership for political decisions as well as critical to good governance (for example, WBGU, 2011).

Fischer (2000) points out the instrumental reasons behind this principle. Since the beginning of the environmental movement, participation in the form of public action and deliberative democracy have been seen as the most effective way to incorporate environmental values into politics (Fischer, 2000, 112). This points to the importance of discursive processes in framing political problems and calling for action. Participation here suggests a way to disrupt dominant power structures that work against sustainability. Therefore, who is involved in the process and how can greatly impact how sustainability is understood and acted upon.

Meadowcroft (2004) distinguishes between three strands of participation relevant to sustainability governance: citizen-based, community-based, and stakeholder-based. Although he argues that decision-makers should use at least two types of engagement, he stresses the importance of stakeholder engagement in promoting deliberation and building partnerships to deal with sustainability issues. By contrast, citizen action is most important in the broader economic, social, and political spheres, with activities such as protest, thoughtful consumption and lifestyle choices, and engagement in civil society organizations and political parties. Meadowcroft's conceptualization of participation in sustainability governance is thus one where different actors and groups are engaged in different levels of social change, from the personal and economic to the political. At the same time, the use of participatory processes in political decision-making functions best with good faith efforts that build partnerships and deliberate in ways that affect problem definition and promote social learning. Therefore, participation is not only inclusion, but must be designed by active governments in a way that affects policy outcomes. It also requires active citizen engagement in public and private.

Not all writers on sustainability are enamored of participatory processes, however. Hajer (1995) points out tensions between expert-driven, top-down processes of implementation advocated by some early writers and consensus-based, bottom-up actions preferred by others. How sustainability, almost universally agreed to be a normative vision, should be defined and implemented thus raises questions of democracy and participation. Related to these concepts is the discussion over forms of knowledge and the role of expertise in policymaking (see also Fischer, 2009; Fischer and Hajer, 1999). These issues are central to sustainability, particularly in politics, because they determine who is expected—or allowed—to define sustainability, and what information may be brought into that process.

At the same time, the literature provides little guidance on how to create and structure participatory processes. What level of effort should be made to include marginalized voices? How should such processes be created, by whom, and to what end? Innes and Booher (2010) warn that truly inclusive decision-making processes are impossible to control; Ehrenfeld (2008) and others discuss the "double-loop learning" of truly deliberative processes, in which outcomes are truly unexpected. Empirical research has also shown mixed results: Healey and Hillier's (1996) analysis of participation in a local planning meeting found that communication was "largely the making of statements rather than free-flowing dialogue" (173). Hawkins and Wang (2011) surveyed U.S. city governments' use of participatory processes and found that participation was mainly defined in passive ways such as information provision (62 percent) and citizen surveys (33 percent), with time-consuming efforts such as consensus-building workshops and roundtables at about 20 percent and less than 5 percent frequency, respectively (17). Participation does not equal deliberation: simply allowing people to speak or listen does not mean that they can influence outcomes. At the same time, little work has been done on the effect such processes have on policy-making. In addition, some worry that use of deliberative mechanisms may reduce the sustainability of outcomes in communities where sustainability values are not widely shared.

At the city level, the principle of deliberation is most clearly evident in the processes used to include different stakeholders and the public in decision-making. For example, development of a strategic plan could use surveys, public meetings, an expert advisory board, or other deliberative mechanisms. Here, it is important to observe three things: who is involved, how the process is structured, and how it impacts policy decisions. Critical here is both the engagement of key stakeholders and citizens, as well as their influence in the decision-making process. Incorporation of this principle in decision-making thus means that the participatory process used impacts the final decisions made. Simply sharing information or holding public meetings is not sufficient.

Reflexive

The systems perspective and understanding of complexity used by sustainability advocates leads to their embrace of the fifth principle discussed here: reflexivity. Described alternately as learning, resilient, flexible, or humble, reflexivity indicates a general approach to decision-making and has been linked to sustainability governance. Reflexivity also fits with deliberation, calling for collection of new knowledge to fit changing circumstances.

Voß, Bauknecht, and Kemp (2006) put forward a model of reflexive governance that draws on Beck's concept of "reflexive modernization." They see reflexive governance as containing two orders of reflexivity. The first, passive, order includes natural feedback loops in governance processes: how the governance system changes to reflect the effects of problem-solving and other interventions (Voß, Bauknecht, and Kemp, 2006, 6). The second order is more mindful, and involves looking back on "its own working, conditions and effects ... interrupt[ing] the automatism of executing problem-solving routines" (Voß, Bauknecht, and Kemp, 2006, 6). Reflexive governance should include five strategic elements: integrated knowledge production; adaptive strategies and institutions; anticipation of long-term efforts of action; iterative participatory goal formulation; and interactive strategy development (Voß, Bauknecht, and Kemp, 2006, 17–20).

Reflexivity is also related to resilience, an increasing focus of literatures on urban sustainability and hazard mitigation (Jerolleman, 2013; Pierce, Budd, and Lovrich, 2011). Resilience refers to how well systems can adapt to change or disturbance. Pierce, Budd, and Lovrich (2011) link resilience attributes to sustainability in U.S. cities. They see resilience, which includes learning, self-organization, and predictive and adaptive capacities, as an important source of sustainability at the urban level. Although their operationalization of resilience includes components more related to political culture and values (for example, with the factor "motivation," as well as social capital and trust in government), they find that it correlates positively with cities' use of sustainability governance instruments and institutions, such as plans, indicators, and sustainability offices.

The notion of reflexivity—listening to other types of knowledge, learning throughout the process, and reflecting on how policy interventions affect complex systems—is evident throughout the literature on sustainability. BUND and Misereor (1996) call for political systems to become more nimble by moving "from tanker to sailboat" (36). Ehrenfeld (2008) speaks of double-loop learning, in which decision-makers learn not only facts and theories to make their decisions, but also learn new ways to understand problems and thus change their goals. Much of the need for reflexivity stems from the necessary humility required to deal with complex systems discussed above.

How does reflexivity manifest itself at the city level? Voß and Kemp's (2006) strategic elements provide some insights. First, agencies and actors should be brought together to deal with problems in an integrated way, rather than in sectoral silos. Here, interagency working groups, advisory councils, and special commissions can help integrate streams of knowledge and build common understanding of policy problems. Adaptive strategies and institutions are constantly re-examined and revised: a strategic plan would be revisited at regular intervals in such a way that stimulated necessary changes to the document. Adaptive institutions would also have mechanisms in place to reduce stagnation and path dependency, such as policies to reduce bias towards political incumbents or sunset provisions that required legislation to be renewed by the legislature after a certain period of time. To anticipate long-term effects, policymakers could use scenario-building techniques that included both likely and less anticipated outcomes in order to forecast and address the city's infrastructure needs. Decision-making processes should be iterative, seeing policymaking as ongoing rather than as one-off decisions to "solve" discrete problems. The extent to which any of these or similar policy instruments and approaches are evident in the six case cities will indicate how well they have incorporated this principle of sustainability into decision-making processes.

Implications for Urban Sustainability Governance

Although sustainability is the subject of intense debate, this chapter has argued that it is guided by five key principles that are generally accepted within the literature. The core feature of sustainability, I argue, is acknowledgment of ecological limits, and in particular the fact that many of these limits have already been exceeded. For this to require urgent action, sustainability works on long time scales, putting a value on the quality of social-ecological systems we leave for future generations. In addition, sustainability challenges the modernist worldview by embracing complexity and systems thinking and encouraging reflexivity in decision-making, which is aware of systemic effects and able to learn and adapt to changing conditions. Finally, acknowledgment of sustainability's standing as a normative proposition that challenges dominant power structures indicates the importance of deliberation and participatory processes that include major stakeholders and the public.

Because of the diversity of the sustainability coalition, different principles are emphasized by different writers, leading to areas of tension and competing priorities despite agreement on core concerns (cf. Kidd, 1992). First, although the environment figures into virtually all writing on sustainability, some writers have emphasized its primacy, focusing on issues of natural limits (to growth and/or resources), ecological carrying capacity, and entropy. Other writers have used this premise to focus on ending growth—particularly economic growth—and how to change our current socioeconomic system to accomplish this. A third, less visible, strand focuses on promoting social equity in a changing society, arguing both that ecological destruction disproportionately harms the poorest and that notions of development much emphasize equity outcomes. Finally, there is a more philosophical—and at times polemical—camp intent on fostering human well-being, by promoting quality-of-life, at times questioning technology, and preaching simplicity and ethical responsibility.

In addition, although the English- and German-language literatures (as well as U.S. and European/German literatures) have not been systematically differentiated in this analysis, one can highlight some key differences. For example, the U.S. literature has seen more interest in complexity thinking and a more technical approach to systems (for example, Meadows, 2008; Espinosa and Walker, 2011), whereas European authors have spent more time dealing with issues of governance (for example, Adger and Jordan, 2009; Kopfmüller et al., 2001; Voß, Bauknecht, and Kemp, 2006) and the social construction of sustainability (Hajer, 1995; Kopfmüller et al., 2001). In addition, some U.S.-based authors have questioned the importance of equity concerns to sustainability (for example, Portney, 2003), whereas European writers draw more openly on socialist and critical theory in defining sustainability (for example, Krueger and Gibbs, 2007). The German and European literatures also include several efforts to deal with international equity (Baatz, 2013) and citizenship issues (Dobson, 2009, 2014; Pettibone, 2014b). Still, these differences are more of emphasis than ideology, as examples can be found from both continents that discuss each of the five principles described here. This cross-pollination is likely increased by the number of transatlantic publications (for example, Krueger and Gibbs, 2007; McCormick et al, 2013; Lafferty and Meadowcroft, 2001; Schreurs, Selin, and VanDeveer, 2009).

In short, the academic literature on sustainability is broad, diverse, and includes some regional differences. It should not be understood as a policy program with specific answers, but a worldview—a pair of glasses that changes how to understand and address social and political problems. To do so, this review has highlighted five key principles that can be seen throughout the literature and linked them to evidence of their incorporation into urban decision-making processes (see Table 2.1).

These five principles of sustainability allow me to operationalize urban sustainability governance in the empirical research that follows as incorporation of these five principles into political decision-making processes. For this reason,

I have given examples throughout this chapter of ways in which these principles could be met in an urban governance context. These principles support the empirical analysis conducted in Part II.

Table 2.1 Linking sustainability principles to urban sustainability governance

Sustainability principle	Evidence in cities
Acknowledges natural limits: Growth cannot continue indefinitely, but is restricted by limits on natural systems. However, the implications for economic growth remain hotly debated.	• Seeks to reduce total resource use rather than increase efficiency • Connects resource use to broader systems in order to understand interactions and effects beyond city borders
Focuses on long time scales: Decisions consider outcomes decades or centuries in the future, not just tomorrow.	• Evaluates policy options on the basis of short- and longer-term outcomes (at least 20 years) • Institutionalizes long-term efforts, inoculating them from the political winds
Uses complexity theory and systems thinking: In contrast to the Cartesian modernist worldview, sustainability sees social-ecological systems as complex, marked by uncertainty, and difficult to understand. It uses a systems view to call for holistic, strategic change.	• Uses strategic instruments such as strategic plans and interagency coordinators to connect policy sectors • Seeks systemic rather than technological change • Includes different types of knowledge to anticipate effects of policy • Acknowledges uncertainty; uses the precautionary principle
Is reflexive: Systems should be redesigned to better accommodate uncertainty and complexity by developing tools to learn and adapt to changing conditions and incorporate various forms of knowledge into decision-making.	• Brings together agencies and actors to build common understanding, through interagency working groups, advisory councils, and special commissions • Revisits policies at regular intervals in such a way that necessary changes can be made • Uses mechanisms that reduce stagnation and path dependency, such as counteracting incumbent bias or including sunset provisions • Anticipates long-term effects with scenario-building techniques that included both likely and less anticipated outcomes and forecast the city's infrastructure needs
Prizes deliberative decision-making processes: Sustainability is a normative concept. Because of this, it must be defined at the local level, in a democratic, consensus-based process.	• Includes different stakeholders and the public in decision-making processes, through surveys, public meetings, an expert advisory board, or other deliberative mechanisms • Uses deliberative inputs in a way that impacts decision-making

Chapter 3
History of Sustainability in the United States and Germany

Efforts to incorporate sustainability principles in the six cities studied here do not take place in a vacuum. This chapter provides necessary context for the case studies at both the international and national levels in the United States and Germany. It opens with a brief overview of sustainability at the international level, highlighting the increasing role of cities in international discussions on sustainability. It then provides national context in both countries under study, focusing on their federal structures, different approaches to urban policy, and divergent action in terms of sustainability. Whereas the United States served as a model for early environmental policy in Germany, its early attempts to take action on sustainability at the national level faltered. Increasing resistance has meant a lack of national framework there. By contrast, sustainability has gained ground in Germany, in part due to several advisory bodies that have framed and advocated for policy decisions to incorporate sustainability concerns. In addition, the European Union has supported sustainability in Germany, as well as providing direct assistance to German cities. The last section of this chapter discusses the important differences and similarities between the two countries in relation to urban sustainability governance.

Sustainability on the International Stage

Sustainability entered the political stage internationally. The United Nations' Conference on the Human Environment in Stockholm in 1972 was the first international conference dedicated to the environment and the largest at the time. It resulted in the Stockholm Declaration, which had 26 principles that deal with improving the environment, promoting social equity, and using or preserving natural resources in a way beneficial to all (UNCHE, 1972). This Declaration laid the groundwork for the UN's work on sustainability, which officially began in 1983 with the creation of the World Commission on Environment and Development, better known as the Brundtland Commission after its chair. Tasked with establishing shared definitions for environmental issues and proposing long-term strategies for international cooperation, the commission published its final report, *Our Common Future*, in 1987 (WCED, 1987). This document coined the term sustainable development, which seeks to reconcile interest in economic growth—particularly relevant to less industrialized countries in the global South—with efforts to address increasing environmental problems (Du Pisani, 2006; Stockdale, 1989). The definition of sustainable development in the Brundtland report is perhaps the most widely cited even today.

Momentum continued to build as world leaders met again in 1992 for the United Nations Conference on Environment and Development (UNCED). Also called the Rio conference or Earth Summit, this meeting led to the creation of five major agreements related to sustainability: the Rio Declaration on Environment and Development, the Convention on Biological Diversity (CBD), the United Nations Framework Convention on Climate Change (UNFCCC), a Statement of Forest Principles, and *Agenda 21* (Lafferty and Meadowcroft, 2001, 13). *Agenda 21* was a critical document that provided the political framework for national, as well as local, action on sustainable development. The agreement, signed by 179 countries including Germany and the United States (Bundesregierung, 2002; Mitra, 2003), contains sections on both the importance of indicators and the role of cities, which it urges to create Local Agenda 21s by 1996 (UNCED, 1992). The document encourages devolution of responsibility "to the lowest level of public authority consistent with effective action" (UNCED, 1992, cited in Dernbach, 2002, p. 10648), singling out local governments as a key player. At the same time, it puts responsibility "first and foremost" with national governments (Dernbach, 2002; UNCED, 1992), situating local action within a national context and calling for the development of national sustainability strategies. A follow-up meeting in 1997 strengthened this call by urging countries to enact their national strategy by 2002. Numerous countries, such as Germany, as well as the European Union, have developed sustainability strategies based in some part on *Agenda 21*.

Since its passage in 1992, implementation of *Agenda 21* has been spotty at best. The follow-up conference Rio+20 in June 2012 was larger than previous UN conferences on environmental issues and included more civil society actors (Biermann, 2013). At the same time, the main agreement, *The Future We Want*, was non-binding and "shrunk to the least common denominator that all countries could support," excluding controversial issues (Biermann, 2013, 1100).

Successive summits have seen the increased engagement of cities working independently of national governments as well as civil society organizations (Bulkeley et al., 2013). Otto-Zimmermann (2012) argues for the importance of including cities in international sustainability governance regimes, and groups such as ICLEI and the C40 (a network of major cities tackling climate issues) have been active at the international level. Indeed, cities' engagement with sustainability is critical, as they are able to act more quickly than is feasible in the international arena. In December 2009, then–New York City mayor Michael Bloomberg went to the climate negotiations in Copenhagen to tout the achievements of his sustainability plan, PlaNYC, and spark international action. But the slow pace made Bloomberg impatient: he and his team left early to get back to work in New York (Bagley and Galucci, 2013, Loc. 914–37). Even if sustainability is an international issue, local action has been more effective, leading urban actors to take an outsized role on the international stage.

Sustainability in the United States

Comparisons of the United States and Germany tend to highlight the two countries' similar economic situation, political structures, and histories of environmental action (Cass, 2007; Kraemer and Schreurs, 2007). However, recent years have seen divergence in policymaking in these two countries recent decades, particularly on environmental issues (Lafferty and Meadowcroft, 2001; Schreurs, Selin, and VanDeveer, 2009). This section begins to explore these differences by presenting the relevant national context in the United States. It makes several key points. First, U.S. federalism and urban policy have both empowered policy action in cities, by granting them relative autonomy and numerous decision-making mechanisms. At the same time, historical bias against urban form and tepid federal support have left cities largely to fend for themselves. National ambivalence and even antipathy towards sustainability have thwarted attempts to develop a sustainability agenda at the national level. This lack of national-level interest has led to development of a loose network of institutions working on sustainability issues at the local level.

U.S. Federalism and Urban Policy

The United States is a federal country that delegates significant political power to subnational units, particularly states and cities. U.S. federalism is enshrined in the Constitution, which devolves power not explicitly enumerated to the federal government to the states. This has allowed U.S. states and municipalities to take the lead in the absence of national action, as seen in the case of climate policy (Rabe, Román, and Dobelis, 2005; Kraemer and Schreurs, 2007). It has also meant a more differentiated policy landscape than in Germany, as can be seen with land-use planning, a key aspect of urban policy.

U.S. federalism has thus left states a larger role in land-use planning activity than is seen in Germany. The U.S. approach stems from the Standard State Zoning Enabling Act passed in the 1920s, which "calls for zoning regulations to be 'made in accordance with a comprehensive plan'" (Light, 1999, 585). The federal government has occasionally intervened in the planning policy, in particular in passing the Federal Housing Acts after World War II, which led to increased suburbanization (Beauregard, 2006; Wheeler, 2004).

Zoning and planning laws in many parts of the United States have historically been used to separate residential from other uses (such as commercial and industrial) and in some cases led to or even sought racial segregation, and tend to encourage homogeneous, automobile-dependent suburban development (Beauregard, 2006; Pallagst, 2007; Rybczynski, 2007). According to Light (1999), this stems from a veneration of private life and a bias on the part of planners against cities, a value also reflected in the Fifth Amendment's language against public use of private property (Wheeler, 2004, 115). Urban form has thus been shaped by the effects of decades of policy that has supported a sprawling, suburban form and left much decision-making to market forces, which Parker

(1997) has called this a "stealth urban policy." Indeed, it is important to note that what key aspects of what is now considered unsustainable in U.S. urban form were intentional features of federal land-use policy.

However, some states and cities have begun to see the negative consequences of such development. The resulting sprawl, loss of agricultural land, and high infrastructure maintenance costs have led Oregon and Washington, among others, to adopt a more active approach to planning. These states have adopted growth management legislation that requires cities and regions to develop plans in order to control land-use growth and channel development into pre-existing areas (Laschever, 1998; Miller, 2004). Portland and Seattle are thus required by state law to create comprehensive plans that actively structure urban form in a way that minimizes urban sprawl. The state-level requirement to develop plans in these states is consistent with planning requirements in Germany (Light, 1999, 582, note 19; see below). Although New York state does not require similar plan development, New York City's geography—which exists of a series of islands that bounds the city on all sides—and development before the passage of zoning laws or popularization of the automobile prevent sprawl within the city's boundaries. Sprawling development thus takes place outside of its jurisdiction and cannot be regulated by city decisions.

The U.S. approach to urban planning has also been charged with preventing sustainable urban development. Urban planners are thus beginning to advocate for new approaches, particularly through smart growth or new urbanism. Smart growth is primarily a response to the problem of "urban sprawl," and prizes development that is compact, with mixed-use development and housing diversity, public and open space, and that is developed in collaboration with local populations (Altrock, Knaap, and Cohen, 2004; Jepson and Edwards, 2010; Kooshian and Winkelman, 2011; Pollard, 2000). New urbanism, developed by architects Duany and Plater-Zyberk, calls for many of the same things, but focuses more on promoting quality of life within smaller, new developments (Altrock, Knaap, and Cohen, 2004; Godschalk, 2004; Jepson and Edwards, 2010; Saha and Paterson, 2008). Both are closely linked to promoting sustainability at the city level and related to urban sustainability in the United States (Godschalk, 2004; Jepson and Edwards, 2010). Thus, cities' understanding of sustainability is closely linked to planners' changing attitudes about urban form.

U.S. federalism has traditionally left urban planning decisions to the state and municipal level, leading to the development of a limited zoning-based planning regime that produced sprawling suburban development. In response, several states have adopted comprehensive planning mandates, requiring municipalities to take a more active role in determining urban form. Two popular planning movements have redefined good planning as pursuit of compact design, mixed-use neighborhoods, amenities, and quality of life; these goals have been linked to sustainability in the city context. Thus, despite historical policies antithetical to urban sustainability and federal inaction, U.S. cities are becoming more amenable to sustainability.

U.S. National Experience with Sustainability

The United States began working on environmental issues in the 1970s and inspired early action in Germany, but the two countries' interest in and approach to environmental and sustainability issues have diverged in recent decades (Kraemer and Schreurs, 2007; Schreurs, 2003). Early U.S. action, led by strong environmental groups (Schreurs, 2003), included the creation of institutions such as the Environmental Protection Agency and the President's Council on Environmental Quality and passage of environmental protection laws related to air and water quality. The 1970s proved a high-water mark for political interest in the environment, however. At the end of the decade, the Carter administration's Council of Environmental Quality released a report on major global trends entitled *Global 2000* that foresees many of the issues detailed decades later in a similar German report (WBGU, 2011). However, the subsequent rise of conservative Republicans led to a backlash against environmental regulations and calls for smaller government (Schreurs, 2003, 2007) and a solidification of the individualist ethos that presents an important challenge to adoption of sustainability principles in U.S. politics (Bryner, 2001; Stedman Jones, 2012).

In 1993, then–U.S. President Clinton created the President's Council on Sustainable Development (PCSD), the only federal institution to date tasked with addressing sustainability. The Council was co-chaired by the president of the World Resources Institute and a vice president of Dow Chemicals and includes 25 high-level presidential appointees[1] tasked with advising the president on issues related to sustainable development (Bryner, 2001; Dernbach, 2002). It defined sustainable development largely in keeping with the principles described in Chapter 2 and the understanding put forth by the Brundtland report:

> Our vision is of a life-sustaining Earth. We are committed to the achievement of a dignified, peaceful, and equitable existence. A sustainable United States will have a growing economy that provides equitable opportunities for satisfying livelihoods and a safe, healthy, high quality of life for current and future generations. Our nation will protect its environment, its natural resource base, and the functions and viability of natural systems on which all life depends. (PCSD, 1996, p. iv)

The PCSD was an advisory council similar to the one developed a decade later in Germany (see below). Over six years, the council gained stakeholder input through numerous public meetings, released three major reports and several working group reports, and briefly introduced sustainability to the national government agenda. It had no formal decision-making authority, but recommended 154 actions in several

1 Kenneth Lay, then CEO of energy corporation Enron, was a member of the PCSD a few years before his company was found guilty of manipulating newly deregulated electricity markets in California in 2001.

policy sectors and 10 goals in its first report (Dernbach, 2002; PCSD, 1996). The local level was singled out as an area of importance:

> Steps toward a more sustainable future include developing community-driven strategic planning and collaborative regional planning; improving community and building design; decreasing sprawl; and creating strong, diversified local economies while increasing jobs and other economic opportunities. (PCSD, 1996, 83)

To Dernbach (2002), the Council's work highlights the real agreement that can be reached on a vision for the United States, even by diverse stakeholder groups (see also Bomberg, 2009, for a positive appraisal of the council's work). Bryner (2001), by contrast, argues that the Council ignored tensions between sustainability and economic growth and "clearly opted for voluntary, incremental efforts aimed at pursuing a modest vision of sustainable development" (288). These evaluations reflect the Council's singularity in attempting to incorporate sustainability into federal policy in the U.S. It remains the only federal government-led attempt to do so and its short lifespan and limited lasting effects highlight the ambivalence towards sustainability in the United States.

The PCSD was ended by executive order in 1999, "because its 'work is completed'" (Dernbach, 2002, fn. 83). Very little evidence of the Council's work remains; it created no new lasting institutions[2] and has not measurably increased interest in sustainability on the national stage. Dernbach (2002) subsequently found 23 unique references to "sustainable development" in the U.S. Code in 2002, three years after the PCSD ended. However, virtually all of the examples he cites focus on implementing or promoting the idea in the so-called developing world, through aid, conservation, and trade programs (see Dernbach, 2002, 10660, fn. 134). Even here, the emphasis is on "economic growth and good governance rather than environmental sustainability, social equity or the need to reconcile these objectives" (Bomberg, 2009, 28). None is clearly related to the PCSD's recommendations.

Although the United States federal government tried to incorporate sustainability into policymaking in the 1990s, the work of the President's Council on Sustainable Development has had little lasting effect. Since then, sustainability has seen rising opposition at the national level. The eight years of the Bush presidency saw a lack of high-level activity on urban planning, environmental, and sustainability issues as well as the continued rise of opposition to such policies within the Republican party (Kraemer and Schreurs, 2007). Local property rights activists joined with the growing Tea Party

2 One positive outcome noted by both Bryner (2001) and Dernbach (2002) is the creation of the Joint Center for Sustainable Communities, with the Council's support and financial assistance, by the U.S. Conference of Mayors and the National Association of Counties. However, according to the Conference's website, the Center was disbanded in 2005.

movement that developed out of the 2008 financial crisis to oppose Local Agenda 21 and sustainability planning specifically (Trapenberg Frick et al., 2014). These groups have focused on state-level action, introducing broad anti-sustainability legislation in 26 U.S. states (Trapenberg Frick et al., 2014). As such, some planners have eschewed use of the terms "sustainability" or "Local Agenda 21" in favor of code words that avoid opponents' attention (Berke and Conroy, 2000; SEA05). Indeed, one of my interview partners in Portland first heard the term "Local Agenda 21" through the opposition movement (PDX05).

Despite conservative opposition, some action has taken place at the national level. In 2009, the Obama administration created the Partnership for Sustainable Communities (PSC), in some ways a reincarnation of the short-lived Joint Center for Sustainable Communities. The Partnership is an interagency collaboration between the Department of Housing and Urban Development (HUD), the Department of Transportation (DOT), and the Environmental Protection Agency (EPA) on urban sustainability. According to its website, the Partnership focuses on reducing sprawling and improving livability (PSC, 2013). Absent is reference to ecosystem health or a global impetus for sustainability. Still, the practical framing of the Partnership's work likely makes it easier to avoid debate on these more controversial aspects of sustainability. More importantly, the PSC links competencies of the three participating agencies related to urban development, allowing a more integrated federal approach to urban sustainability. It distributes federal grant monies to, for example, public transportation, affordable housing and mixed-use development, and brownfield redevelopment projects.

Despite the void in federal government activity related to sustainability, several non-governmental organizations are active at the national level on urban and sustainability issues in the United States. Although there is significant overlap, these organizations fall into three major categories: groups focused on urban issues, environmental nonprofits, and organizations working explicitly on sustainability. Each provides some level of support to the advancement of sustainability in U.S. cities.

Major current players on urban issues at the national level include the National League of Cities (NLC), Smart Growth America, the Urban Institute, the American Planning Association (APA), the Council for the New Urbanism (CNU), and the U.S. Green Building Council (USGBC). In lobbying for city-friendly policies like public transportation funding (Sheehan, 2001) and changes in federal real estate policies (Smart Growth America, 2013), these organizations often use the language of sustainability to support their arguments and describe their vision. The planning-oriented groups have done important work to develop new urban planning frameworks that are more compatible with sustainability principles, such as smart growth and new urbanism. For example, the APA created the Sustaining Places Task Force in 2010 to demonstrate the role of the comprehensive plan "as the leading policy document and tool to help communities of all sizes achieve sustainable outcomes" as well as to highlight best practices for sustainable cities (Godschalk and Anderson, 2012, 1). The task force's 2012 report presents eight principles to be used to incorporate sustainability into comprehensive planning. These groups have developed new principles of good planning and sustainability at the urban level, and advocated for their implementation by urban planners and city decision-makers.

Traditional national environmental nonprofits like Environmental Defense Fund (EDF), the Sierra Club, and the Natural Resources Defense Council (NRDC) have done less work related to either sustainability or cities, although the NRDC does run a Smarter Cities program, which publishes information on how urban planning and form relate to positive environmental outcomes. Indeed, a study on the failure to pass national cap-and-trade legislation in 2009 pointed to poor strategic decisions by these groups, including lack of public outreach (Skocpol, 2013). Much has been written about the anemic environmental movement in the United States (for example, Brulle, 2000); environmental nonprofits have not only failed to spark a large-scale environmental consciousness, they have generally not focused on a more holistic sustainability perspective. Still, as will be seen in the New York case, the regional offices of these groups have helped steer sustainability at the local level.

A third type of group is relatively new and focuses explicitly on sustainability, often at the local level. Here ICLEI USA—the U.S. branch of the international NGO dedicated to "local governments for sustainability"—is the most visible example, but other groups include the Resource Renewal Institute, U.S. Green Building Council, SustainLane, the Center for a New American Dream, and Redefining Progress. The diverse approaches of these organizations, from green building to transportation to sustainable consumption and indicator work highlights the richness of the U.S. nonprofit landscape (see also McGann and Johnson, 2005). The U.S. Green Building Council developed the architectural sustainability standard LEED, which, although criticized for the perverse incentives of its point system (for example, Owen, 2009), has given developers tools to measure sustainability in the built environment.

A closer look at these groups, however, reveals the ambivalence of the American public to sustainability. ICLEI USA has been under constant attack by a small group of conservative activists since its creation in 1993. SustainLane stopped publishing its sustainable city rankings in 2008 and Redefining Progress, creator of the Genuine Progress Indicator (an alternative to GDP), has reduced its profile since its high-level critique of GDP in the 1990s (Cobb, Halstead, and Rowe, 1995). Although these organizations do important work, they have not managed to create a national-level discussion on sustainability. In addition, particularly the work of groups focused on urban sustainability can be seen as an effort to counteract suburbanization in the United States, as well as strengthen the relatively weak urbanist movement. Although lobbying and civil society organizations have been said to be more important in the U.S. than Germany (McGann and Johnson, 2005), there is a limited number of national groups seeking to influence sustainability in U.S. cities.

This section has revealed a mixed history with sustainability in the United States. This is reflective of the country's troubled experience with environmental and social movements, which have challenged its individualist ethos and questioned neoliberalism (Bryner, 2001; Stedman Jones, 2012). This history is evident in the country's ambivalent attempts to implement sustainable development at the national level, the relative weakness of its national-level organizations related to sustainability, and the strong resistance to sustainability among a small minority of conservatives. These factors make it difficult to implement national-level initiatives based on sustainability, but also make it possible for cities where fewer of these factors are present (such as in environmentalist havens like Seattle and Portland) to be more active in taking policy action.

Since the brief existence of the President's Council on Sustainable Development in the 1990s, there has been no high-profile national political work to incorporate sustainability into decision-making. Individual programs exist in several federal agencies, but these programs do not provide a comprehensive framework necessary to understand urban sustainability initiatives in U.S. cities. These programs implement individual projects and seek to reduce anti-urban and anti-sustainability bias in federal decision-making rather than to demonstrate a serious commitment to urban sustainability at the national level. Since the political deadlock that followed Barack Obama's election and the rise of the Tea Party, some have expressed increasing doubts in government's ability to solve major policy problems (Ornstein, 2013; Skocpol, 2013). Although not the subject of this book, it is important to note the possible "unsustainability" of a political system that responds to crisis by creating larger problems of its own. If sustainability is concerned with the resilience of public institutions—that is, their ability to deal with wicked problems and unexpected events—then there is much work to be done to make the U.S. federal political system more sustainable and thus better able to solve rather than create problems.

Table 3.1 Sustainability in key U.S. publications

Publication title	Year	Author, Institution	Main points	Link to sustainability principles
Global 2000	1980	Council of Environmental Quality	• Calls on federal government to project long-term implications related to population, natural resource, and environmental trends • Call for continuous review of federal agency assumptions and assessment methods	• Acknowledgment of limits • Long time scales • Iterative, reflexive approach
Sustainable America: A new consensus for the future	1996, 1997	President's Council on Sustainable Development (PCSD)	• Calls for decoupling economic growth from environmental harms • Emphasizes equality of opportunity • Stresses the importance of consensus on long-term vision and goals • Calls for individual responsibility and stewardship	• Acknowledgment of limits • Calls for integration of policy issues • Deliberative approach within Council; engagement with public, calls for active citizens
Our common journey	1999	National Research Council's (NRC) Board on Sustainable Development	• Transition will take place over next two generations • Must meet needs of larger, stabilizing population • Must support planetary life support systems • Must reduce hunger, poverty	• Acknowledgment of limits • Long time scales
Sustaining places: The role of the comprehensive plan	2012	American Planning Association's (APA) Sustaining Places Task Force	• Sees comprehensive plan as key instrument to implement sustainability in cities • Stresses the importance of "authentic participation," "harmony with nature," equity, and economic resilience	• Link to regional plans reflects systems thinking • Calls for deliberative plan development

Sources: In addition to the primary sources, several academic analyses were consulted. For *Global 2000:* Dearborn, 1983. For *Sustainable America*: Bryner, 2001; Dernbach, 2002. For *Our Common Journey*: T. Miller, 2011.

Sustainability has little national presence in the United States today. This may allow cities interested in sustainability more latitude in defining its principles and related policies, but also shows that city action is voluntary, and takes place in a national context that could be considered ambivalent to sustainability.

Sustainability in Germany

The German context for urban sustainability differs from that in the United States in several respects. This section shows that, although Germany took up the cause of sustainability later than the United States, its incorporation into policymaking at all levels has been aided by 1) the development of institutions working on sustainability issues, 2) an expert-led understanding of sustainability, and 3) the material and institutional support of the European Union. In addition, sustainability is more compatible with long-standing German values of solidarity and newer political concepts such as the precautionary principle.

German Federalism and Urban Policy

German federalism is based on the principle of subsidiarity, in which government action is to be taken at the most suitable level (Kraemer and Schreurs, 2007). This has led to a more functional form of federalism than in the United States, as states (*Bundesländer*) are generally tasked with implementation of federal policies (Jörgensen, 2012). However, the German system provides states and cities less power to develop policy, partly because of a more comprehensive regulatory framework on policy issues affecting cities (Jörgensen, 2012; Light, 1999). This policy framework seeks to unify integrated land-use policy across jurisdictions and promotes a positive vision of cities as a critical part of metropolitan economies (Light, 1999).

German urban policy, in contrast to the United States, seeks to reduce sprawl, protect traditional urban and rural forms, and prevent large-scale migration within the country. A central feature of German planning law is the designation system of towns and cities that permits certain kinds of development in different types of areas—such as "central cities" and "small centers"—and preserves the centrality of established major cities (Larsen, 1996; Light, 1999). This system sees regions as interconnected systems of services and infrastructure, allowing the government to "[channel] certain types or levels of development and limited government resources into particular areas, and in so doing, attempts to assure that all citizens have access to a full array of work, social, cultural, and educational opportunities within a reasonable distance of their homes" (Larsen, 1996, 1007).

German cities have thus been relatively stable over the past 50 years. The country has seen less of the rapid urban growth or decline that marks most U.S. cities

(Larsen, 1996).[3] Integrated German policy has led to strong transportation links throughout the country as well as robust public transit networks within cities. Although these policies have been watered down since German reunification in the 1990s, they have helped maintain compact, well-connected, healthy mid-sized cities throughout the country.

Urban planning in Germany includes detailed spatial plans at all levels of government. The German *Bundesraumordnungsgesetz* (Federal Land-Use Law, or BROG) sets land-use goals and requires states to develop multi-level land-use plans (Light, 1999; Scholl, Elgendy, and Nollert, 2007). Light (1999) argues that this distinction is due to a vision of the city as a necessary part of social life to be managed by the state. Light summarizes this difference between how the two countries approach planning:

> Thus, an important conclusion … is that our land-use regime must be seen in the context of broader American attitudes toward the relation between individual and community. If German and Swiss land-use regulations place the individual in the center of a managed landscape and a restrictive community, American land-use regulation seems to reject this idea of a socially situated self. If German and Swiss land-use policies reflect a dogged determination to control the modern town and make it conform to an inherited idea of the well-made community, American land-use controls represent a willingness to experiment with unprecedented forms of urban life. (611)

Germany's more hands-on, managed approach has led to several differences in urban form from the United States. First, cities enjoy a privileged status: they are seen as part of an integrated system that includes urban and rural areas; thus policies seek to maintain their health and vitality (Light, 1999). This is done by policies that both reduce competition between cities but also seek to preserve urban cores by promoting mixed uses, integrated transportation links, and compact development. Second, German regulation does not seek to separate single-family homes from other uses nor segregate land-use purposes in general, as seen in the U.S. approach. Light (1999) argues that this is due to different underlying myths of the city, a notion in keeping with more general writings on the key difference between the United States and Germany: namely, how each values the individual vs. the community. German policy reflects a cultural interest in preserving healthy communities, which has led to a more hands-on interest in urban planning. The focus on the individual in the United States has led to more sprawling development

---

3 Although the German law on spatial planning (BROG) reduced movement in general, it did not eliminate the migration of one million residents of the former East Germany to the West after reunification. Still, West German policies were seen as important to increase stability in its former neighbor and have since been modified to add the goal of "spatial integration" between the two countries. Planners have used the BROG to address sustainable development as well (Larsen, 1996).

that prizes single-family homeownership and separation of the family unit from the bad influence of the city.

To protect urban form, German law requires lower levels of government to develop certain spatial plans that guide urban development in accordance with federal law. Most important to the city are land-use plans (*Bebauungspläne*) and zoning plans (*Flächennutzungspläne*), which designate planned uses for individual parcels of land within the city at increasing level of detail (Scholl, Elgendy, and Nollert, 2007). These are required by law to match the state development plan (*Landesentwicklungsplan*) (Scholl, Elgendy, and Nollert, 2007).

Bückmann (2002), analyzing German law's compatibility with sustainability goals, finds federal planning law to be the furthest along. Sustainability is an explicit goal in federal building law (*Baugesetzbuch*) (Bückmann, 2002, 156) and land-use planning law (BROG) (Gehrlein, 2004, 97). Important in both laws is the call for economic, social, and environmental needs to be harmonized. However, Bückmann (2002) finds no practical consequences to the legal embeddedness of sustainability in German land-use law.

Beyond the role of sustainability, the influence of traditional legal planning instruments has nevertheless declined as private landowners and users exert more influence. To deal with this and develop more strategic policymaking instruments, cities have begun using strategic development plans (Gehrlein, 2004, 89ff.). Usually in the form of urban development plans (*Stadtentwicklungspläne*) or vision documents (*Leitbilder*), these new plans do not simply order the built environment, but set general rules for planning based on a strategic vision of the city. Because strategic plans are not part of the federally mandated planning frameork, however, they have seen uneven implementation. As such, although these plans can play a guiding role in setting a unified strategy, they often serve as partisan political visions that are ignored in contentious planning decisions or by incoming administrations. Their lack of legal standing makes German strategic plans roughly comparable to the comprehensive plans passed in U.S. cities.

In addition to strategic plans, Agenda 21 has played a larger role in Germany and the European Union than in the United States, causing many German municipalities to create Local Agenda 21s (LA21s) to incorporate sustainability into city decision-making. LA21s can also provide a strategic framework for urban decision-making. However, as the local implementation tool of the UN sustainable development framework *Agenda 21*, LA21s also explicitly call for plans to use sustainability principles and be developed in a participatory manner. This seems to make them the ideal tool to incorporate sustainability principles into urban decision-making. German experience with LA21s has been uneven, however. German LA21s have rarely served as a holistic planning instrument but instead have focused on individual projects or programs, often linked to only one or two aspects of sustainability (Beuermann, 1998; Kern, Koll, and Schophaus, 2007). However, supralocal pressure on cities to develop a Local Agenda 21 has created momentum and the financial resources in some cases to develop a document consistent with strategic planning. All three of the German cases studied here have created some form of Local Agenda 21, although these initiatives have had a limited effect on incorporating sustainability into the urban decision-making process.

In short, German federalism and the country's cultural context have created an atmosphere more conducive to urban planning and more supportive of cities. Federal laws require that cities create and regularly update plans that preserve urban form and promote specific land uses at the parcel level. In addition, Germany's higher level of engagement with Agenda 21 has seen the development of Local Agenda 21s in cities. However, the empirical experience with LA21s and strategic planning has been uneven. Thus, although Germany has background conditions more amenable to strategic planning for sustainability, its cities have not always gone beyond the traditional planning tools available to them to develop more overarching strategic plans.

German National Experience with Sustainability

Germany's experience with sustainability contrasts sharply from that of the United States. For a European country, Germany began developing its national sustainability strategy relatively late. Since then, however, sustainability has been supported more consistently: by major expert publications, civil society interest in sustainability, and the support of the European Union (EU). The last decade has seen rising interest in environmental and sustainability issues in Germany. The most recent of these is the energy transition (*Energiewende*), a high-level national initiative to transition the German electricity mix from nuclear and coal towards renewable energy in the coming decade (Morris and Pehnt, 2012). In addition, sustainability-related issues have not taken on partisan flavor seen in the United States (Schreurs, 2003); center-right CDU chancellor Angela Merkel has pushed numerous environmental initiatives and supports sustainability institutions.

Environmental policy in Germany began with the U.S. model, but was further advanced by the green grassroots and the Green party, which entered the Bundestag in 1983. Beuermann (2001) argues that the importance of the precautionary principle in Germany was one reason why the country ignored sustainability for so long (91). However, she finds that sustainable development narratives began to supplant—or at least supplement—climate change in German environmental policy by the beginning of the new millennium. This may be due in part to its compatibility with the German social market philosophy, which gives the government a greater role in promoting addressing market failures and promoting equity (Schreurs, 2003, 11).

The increasing salience of sustainability in Germany can be seen in the proliferation of expert institutions and reports related to sustainability, Germany's national sustainability strategy, the number of actors working on these issues today, and the role that national-level engagement plays in German cities. Sustainability has been defined in Germany through complex interactions between government, research institutions, and advocacy groups. Unlike in the United States, several national-level organizations in Germany have spent a great deal of time in the past two decades trying to define sustainability and develop frameworks for action. Several key reports came from organizations with close ties to the government or the Bundestag. Indeed, the sheer scope of work at the national level in Germany makes it impossible to cover everything. I thus aim to give a general sense of national efforts and provide more detail on actions and research relevant to cities. Table 3.2 presents an overview of key publications related to sustainability in Germany.

Several books have been written on the state of sustainability in Germany that include concrete actions for improvement. The first report on sustainability written by an institution linked to the federal government was that of the German Advisory Council on the Environment (SRU) in 1994 (Brand and Jochum, 2000, 43; SRU, 1994). The council wrote that environmental policy needed an integrative approach to understand interactions between diverse issues and chose sustainable development—or literally "lasting, environmentally just development"—as an overarching vision. Its model explicitly criticizes the growth vision present even in the Brundtland definition (Brand and Jochum, 2000, 45). At the same time, the SRU report's call for change appears to derive primarily from individuals' "voluntary acceptance of responsibility" and agreement to operate within ecological limits (Brand and Jochum, 2000, 64).

In 1996, environmental group BUND co-produced *Sustainable Germany* with Misereor, Catholic development organization, with the assistance of the Wuppertal Institute (BUND and Misereor, 1996; Zahrnt, 2014). BUND and Misereor (1996) provide three scenarios to suggest possible futures for Germany: business as usual, an offensive energy strategy to improve efficiency and reduce climate emissions and pollution, and a "sustainable Germany" option that would require additional lifestyle changes. *Sustainable Germany* makes several important points about sustainability as a transformational process. They define sustainability as moving "from tanker to sailboat," creating a system that is more mobile and resilient to change (BUND and Misereor, 1996, 36). Instead of a capital approach, the authors use the concept of environmental space, which includes "international justice, resilience, and freedom" (BUND and Misereor, 1996, 27–8). Their recommendations to eliminate environmentally damaging subsidies and prioritize the good life over consumption (BUND and Misereor, 1996, 206) have been generally accepted by advocates in the last decade

The book's main function is visionary: contextualizing major policy problems in the language of sustainability and creating a vision for a desirable sustainable future. It formed an important foundation for sustainability discussions in Germany; according to Beuermann (2001), this "probably had the most significant impact in stimulating public debate on sustainable development" (92; see also Brand and Jochum, 2000). It was succeeded by a follow-up study by Coenen and Grunwald in 2003, which used scenario analysis to evaluate sustainability. Although it is difficult to draw an explicit line between these early reports and subsequent political action, this debate has highlighted issues in the ongoing

Table 3.2 Sustainability in key German publications

Publication title	Year	Author, Institution	Main points	Link to sustainability principles[a]
Environmental Assessment 1994: For a Lasting, Environmentally just development	1994	German Advisory Council on the Environment (SRU)	• sustainable development used as an "integrated approach to environmental policy" • questions growth and calls for recognition of human connection to natural systems	• emphasis on limits and long time scales in definition of sustainable development • call for integrated approach uses complexity
Sustainable Germany (Zukunftsfähiges Deutschland)	1996	German Friends of the Environment (BUND) and Misereor	• cooperation between major environmental NGO and a Catholic development organization significantly increased • public debate on sustainable development • calls for flexible governance • uses environmental room (*Umweltraum*) instead of a capital approach • stresses the good life over material wealth	• discussion of "environmental room" consistent with limits • desired move "from tanker to sailboat" suggests reflexivity
Concept for Sustainability: From Model to Implementation	1994, 1998	Enquete Commissions	• focuses on developing a long-term strategy for resource use • internally inconsistent in how to deal with trade-offs	• acknowledges limits • long-term • deliberative in approach: developed through broad political discussion
National sustainable development strategy	2002, 2004, 2008	Federal government	• international and intergenerational responsibility as thematic area • sees sustainability as "guiding principle"	• addresses long time scales • efficiency goals at odds with notion of absolute limits • public consultation reflects deliberative approach

Sustainability Problems in Germany	2003	Coenen and Grunwald	• defines sustainability as equitable, integrative approach to future based on scientific knowledge • calls for development of key technologies for sustainability transition • argues that change is required in relationship between politics, science, and society	• acknowledges long-term, complexity • acknowledges need for deliberation
Social Contract for Sustainability	2011	German Advisory Council on Global Change (WBGU)	• frames transition to sustainability in terms of values change • calls for massive macro-level policy and governance changes	• long time scales understood: transition on generational scale • deliberative principle not fully understood: public seen as passive actor
Cities for a Sustainable Germany	2011	German Council for Sustainable Development (RNE)	• sustainability as strategic orientation to unify policymaking • deals with financing, participation, and monitoring	• acknowledges complexity in call for policy integration • includes deliberation • calls for reflexivity
Responsibility in a Finite World	2012	German Advisory Council on the Environment (SRU)	• environmental limits are being exceeded • discusses need to decouple resource use from prosperity • argues for valuation of ecosystem services • calls for integrative political approaches	• acknowledges ecological limits • uses complexity

Sources: In addition to the primary sources, the following academic analyses were consulted. For SRU 1994 and Enquete Kommission 1994: Brand and Jochum, 2000. For BUND and Misereor 1996: Zahrnt, 2014. For Bundesregierung 2002: Leukhardt and Allen, 2012.

Note: [a] See Chapter 2 for discussion of sustainability principles used throughout this book.

national discussion related to sustainability, namely attempts to support qualitative well-being over quantitative economic growth, debate on the need for massive lifestyle change, and questions surrounding the appropriate role of technology in transformation.

In addition to these books, the German Bundestag has called several Enquete Commissions on major questions related to sustainability. The Commission on Protection of Man and the Environment released its final report, *Shaping industrial society*, in 1994 to create a clear vision to address the environmental problems caused by the chemical industry (Brand and Jochum, 2000, 68ff). A follow-up commission developed a more comprehensive vision of sustainability, which was released as *Concept for Sustainability: From Model to Implementation* in 1998 (Enquete Kommission, 1998). More recently, the Enquete Commission on Growth, Wellbeing and Quality of Life – Paths to Sustainable Economic Activity and Social Progress in the Social Market Economy, created in 2010, was created by the SPD and Greens to develop an alternative vision to GDP growth.[4] The commission failed to develop a single document with support across political parties, highlighting the contentiousness and intensity of the growth debate in Germany (Pinzler, 2013). Still, members of the commission are hopeful about the fact that questioning of the growth paradigm was happening within—rather than between—the major political parties (personal communication, 2013).

Scientific advisory bodies continue to release reports to conceptualize change required at the international, national, and local level. The German Advisory Council on Global Change (WBGU) released its *Social Contract for Sustainability* in 2011, calling for massive decarbonization of the global economy by 2050 based on its earlier carbon budget concept (WBGU, 2005). The report mentions urbanization as one of the three main areas for transformation, but focuses mainly on climate change and centralized or global solutions in its recommendations. As an internationally focused body, it may not be surprising that its calls for more participation are not expanded to include local level innovations or community-based solutions.

Perhaps the most relevant expert publication for urban sustainability, the German Council for Sustainable Development (RNE) released a report in 2011 based on its dialogue of mayors on sustainable cities that focuses on sustainability in German cities (Grabow et al., 2011). This document, *Cities for a Sustainable Germany*, presents a comprehensive understanding of sustainability in German cities. It conceptualizes sustainability as a strategic orientation that allows cities to unify their policymaking. The report includes important issues for German cities, such as financing, participation, and monitoring. Understanding sustainability as an overarching approach, the report highlights "municipal strategies and steering and management approaches that allow sustainability goals to be reached" (4).

The sheer number of reports released by German government and expert groups, and the breadth of areas they discuss, shows sustainability's political

4 Anke Hassel event, 2013.

salience in Germany as well as the key role played by experts in calling for action. One key action at the national level is the country's national sustainability strategy. Although the strategy has been reviewed critically by some scholars, it represents a serious attempt to begin to use sustainability as a guiding principle for policymaking.

Germany first published its national sustainability strategy in 2002 (Bundesregierung, 2002); since then, it has released two progress reports (Bundesregierung, 2004; Bundesregierung, 2008) and two indicator reports (Destatis, 2007; Destatis, 2008). In addition, the Chancellor's office and the German Council for Sustainable Development called for a peer review of its sustainability strategy in 2009 and 2013. The resulting reports read more as cheerleading than critical reviews, however, calling for Germany to brand itself as a green economy and take a stronger leadership role in the European Union on sustainability issues (RNE, 2013; Stigson et al., 2009).

Leukhardt and Allen (2012) take a critical stance on the German sustainability strategy. They find the thematic areas somewhat arbitrary, the approach to environmental concerns piecemeal (for example, forests and water pollution are not included), and several of the strategy's goals at odds with sustainability principles. Leukhardt and Allen (2012) also question the strategy's use of GDP as a measure of economic prosperity, citing concerns raised by Daly and Cobb (1989) and Jackson (2009). The differences between these reviews highlight the authors' different frames of reference. The official peer review compares Germany to other countries and sees the glass as half-full: Germany's strategy is a model for Europe and the world. Leukhardt and Allen's analysis draws from sustainability principles, leading them to see the glass as half-empty: the strategy does not fully reflect sustainability principles. Both assessments are likely true, indicating the disappointing realities of political efforts on sustainability.

Many German cities cite the national and EU strategies in their own sustainability documents, and a former member of the German sustainability council referred to the national strategy as setting the reference standards for local efforts (DE01). Only one of the strategy's national goals—land conversion for housing and transportation—is explicitly related to the local level. However, this goal has been tied to national planning, building, and taxation instruments, for example, property tax reform that reduces suburbanization and promote urban brownfield redevelopment (Bundesregierung, 2002, 287ff). The national strategy actively engaged input from the states, many of which have incorporated the federal sustainable development strategy into state-level plans (Niestroy, 2005). However, the national strategy seems to use a voluntary approach to lower-level engagement: "the strategy has not triggered any new Land into becoming active, and also does not call upon the Laender to prepare their own SDSs [sustainable development strategies]" (Niestroy, 2005, 141). This suggests that those states not previously engaged in sustainability issues have not been urged to become so by the national framework.

The role of motivating policy action in Germany thus falls to the numerous institutions focused on sustainability and related issues, many of which we created to help develop and coordinate sustainability in national politics. For this reason, these groups have a closer relationship to national policymakers than is true in the United States. In addition, the landscape of sustainability groups in Germany is far more academically or research-oriented than is true in the United States, where sustainability groups focus more on networking or advocacy efforts, or discrete parts of sustainability such as smart growth. This section highlights the work of three types of groups: expert advisory councils, government sustainability networks, and external nonprofit groups.

Germany relies more on expert advisory councils for policymaking than the United States, including several groups focused on environmental and sustainability issues. The most important of these is the German Council for Sustainable Development (*Rat für Nachhaltige Entwicklung*, or RNE). Created in 2001, in part to aid development of the federal sustainable development strategy, the 15-member council includes two mayors; heads of a major environmental organization and the Chamber of Commerce; former environment ministers; and civil society leaders (Niestroy, 2005). The council's three main duties are to provide input on the national sustainability strategy, name concrete issue areas and projects, and strengthen the public discussion on sustainability (DE01). It publishes research and conducts projects related to different aspects of sustainability; advises the government on indicators, selection of overarching themes, and pilot projects; and issues statements on government reports. According to Niestroy's (2005) interviews, the German sustainability council is generally seen by government actors as a watchdog, think tank, stakeholder/expert dialogue, agenda-setter, advocate for a holistic view of sustainability, and promoter of outside dialogue. The council has done important work on urban sustainability, particularly through its sustainable city dialogue with mayors of major cities, which has published several reports on how sustainability can be realized at the city level (Grabow et al., 2011; Werner and Dorfner, 2013). These reports serve essentially as handbooks for mayors who want to start their own process to incorporate sustainability into their city (DE01), highlighting the council's critical role in promoting sustainability at the local level.

In addition to the sustainability council, the federal government set up the German Advisory Council on Global Change (*Wissenschaftlicher Beirat der Bundesregierung Globale Umweltänderungen*, or WBGU) in 1992 in the run-up to the Rio Earth Summit. The WBGU has nine members, appointed for a term of four years by the federal cabinet, is supported by an interministerial committee that represents the federal ministries and chancellery, and serves as an independent, scientific advisory body to the government. The Council's principal task is to provide the government scientifically based policy advice on global change issues (WBGU, 2011). Whereas the sustainability council also works closely with local and state-level actors to implement national policy on sustainability, the WBGU looks at the federal and international level and provides a more high-level, long-term vision of changes to be made, as can be seen in its 2011 report on the "social contract for sustainability." At the same time, the group's academic focus gives it a more theoretical bent, as seen in that report's rational model of the policy process.

In addition to advisory councils dedicated to sustainability and global change, the German government relies on the German Advisory Council on the Environment (*Sachverständigenrat für Umweltfragen*, or SRU), created in 1971, to provide the federal ministry of the interior and later the environment ministry expert advice on environmental issues (Brand and Jochum, 2000, 43). The SRU writes academic reviews of major environmental topics commissioned by members. It releases an assessment report (*Gutachten*) every four years (formerly every two) on a major environmental issue, and has often discussed issues related to sustainability. The 1994, 1996, and 1998 reports focused on defining "the concept and implementation of sustainable development" (Beuermann, 2001, 96); the 2012 report is entitled "Responsibility in a finite world" (*Verantwortung in einer begrenzten Welt*) and focuses on three issues: decoupling resource use from economic prosperity, valuation of ecosystem services, and strengthening integrative concepts (SRU, 2012). In this way, it attempts to weigh in on the current growth debate in German politics and connect to the related Enquete Commission. Generally, SRU reports provide an expert underpinning for political decisions on complex environmental issues by framing environmental problems and suggesting appropriate targets and thresholds. The seriousness with which policymakers use SRU reports highlights the greater reverence for scientific information and the less contested nature of environmental concerns in the German context.

In addition to these expert advisory bodies, the national level of German government includes several committees and networks organized to coordinate action and exchange information. In the executive branch, the State Secretaries' Committee on Sustainable Development, or green Cabinet, is a group of nine state secretaries[5] from the major federal ministries (Niestroy, 2005, 138), and is primarily responsible for monitoring and implementing aspects of the national sustainability strategy related to their purview (Stigson et al., 2009). The German Ministry for Education and Research (BMBF) has developed a socio-ecological research program that funds research related to sustainability transition and provided funding instrumental in developing Voß et al.'s 2006 volume on reflexive governance for sustainability (xvii).

In the legislative branch, the Parliamentary Advisory Council on Sustainable Development (*Parlamentarische Beirat für nachhaltige Entwicklung* or PBNE) was created in 2004 (and renewed in 2010) on the recommendation of the sustainability council to assist the legislative implementation of German and EU sustainable development strategies, by providing legislative points of contact for sustainability actors (Niestroy, 2005, 140). Unlike normal committees, this

5 State secretaries (*Staatssekretäre*) are high-level staffers within the ministries. "*Beamtete*" secretaries are administrative staffers, while "*parlamentarische*" secretaries are members of the Bundestag who serve as politically appointed ministry undersecretaries.

22-member body must be reconstituted each electoral period; it calls for debates, works with standing committees, and serves as a watchdog, but is not allowed to make recommendations (Niestroy, 2005, 140). These groups demonstrate the serious interest in implementing the federal sustainability strategy at the national level.

Finally, non-governmental groups are engaged in sustainability at the national and subnational levels. In particular, BUND, the German branch of the environmental advocacy group Friends of the Earth, released an early major report on sustainability in Germany titled *Zukunftsfähiges Deutschland* (Sustainable Germany; BUND and Misereor, 1996). As discussed above, this publication helped to raise interest in sustainability in the late 1990s. As will be seen in the case studies, local chapters of environmental groups such as BUND and NABU have done important work defining sustainability and developing projects to implement it in the case cities.

Germany and the European Union

In addition to the more consistent, active engagement of the German federal government on sustainability issues, Germany's membership in the European Union provides additional support for German cities interested in transitioning to urban sustainability governance. The EU affects German cities in three ways: through positive peer pressure and EU-level interest in and policies towards sustainability, the use of networks that provide resources and information, and financial instruments that aid cities' implementation of sustainability initiatives.

The EU has allowed member states such as Germany to engage in an environmental race to the top, but has been criticized for its continued pursuit of growth. Germany has pushed hard for EU-level policies related to sustainable development and included the theme of integrated urban development in its 2007 EU presidency. The EU also negotiates a leader position at UN conferences related to sustainable development and climate change, pushing countries like the United States toward bolder action and seeking to shame them with first-mover activities (Bomberg, 2009). In 1998, the Cardiff Process called for integration of sustainability and environmental objectives into diverse policy sectors (Bomberg, 2009, 25). The EU created a sustainable development strategy in 2001; sustainability is a key feature of EU research and program funding (Bomberg, 2009; Zimmerman, 2007). However, the EU's definition of sustainability has raised concerns with some analysts. Peters (2003) finds that EU documents use an ecological modernization framework that allows economic growth and sustainable development to exist without tension. Bomberg's (2009) and Pfister and Flear's (2014) more recent analyses come to similar conclusions. These authors argue that the EU's sustainable development strategy, conducted in parallel to an unchanged economic growth agenda, does not actually prioritize sustainability issues. Concern for the environment becomes another component to consider in an otherwise stable regime that does not address the larger problems posed by sustainability. Bomberg (2009) also discusses a reduction in emphasis on sustainability in EU documents after 2002 (26). Thus, although the EU has sought to include sustainability as a guiding principle, it is unclear the extent to which action represents true change or understanding of sustainability as connected to the status quo.

The second way in which the European Union supports sustainability in German is through European-wide networks that provide peer pressure, resources, and information related to sustainability. For example, the 1994 Aalborg Charter includes a policy framework for creation of local plans, establishes the European Sustainable Cities and Towns Campaign to encourage further local activity, and calls upon cities to establish Local Agenda 21 processes (Van Begin, 2004). The Charter now includes over 2,600 signatories, including all three German cases studied here, according to the Aalborg website (2013). The more ambitious Aalborg Commitments created in 2004 require cities to engage in stakeholder dialogues to set targets to be incorporated into LA21s or other sustainability action plans. Signatories are asked to send information on their targets and achievements to the Campaign in order to allow regional monitoring and benchmarking (Van Begin, 2004, 216). Hamburg and Heidelberg are two of the nine German cities that have signed the commitments (of 666 cities throughout Europe).

Another European-wide network is EUROCITIES, which, according to its website (2013), lobbies for urban interests in European policymaking. EUROCITIES has several working groups and forums to help cities with sectoral issues related to sustainability, and includes climate and inclusion goals. Both Berlin and Hamburg are full members of the network. Finally, the Covenant of Mayors is a group founded in 2008 that coordinates local efforts on climate change, particularly related to energy efficiency measures. A voluntary instrument, the Covenant helps cities to develop ambitious climate action plans to reduce emissions.

A third way the European Union supports German cities is by providing resources. This takes the form of institutional support such as research and pilot projects, as well as financial resources for pilot projects, research, and capacity building. The EU provides several programs and initiatives explicitly focused on urban sustainability. For example, the first four European Conferences for Sustainable Cities and Towns were funded by the European Commission (Zimmermann, 2007). The affiliated European Sustainable Cities and Towns Campaign was created in 1994 and to help mainstream sustainability best practices and implement the Aalborg Charter and Commitments in European cities. Heidelberg won the Campaign's Sustainable City Award in 1997 and 2003, and Hamburg was awarded European Green Capital in 2011, which provided financial resources to conduct a citywide sustainability study.

The EU has also conducted several pilot programs to build best practices, such as Managing Urban Europe 25 (MUE-25), a pilot program in 25 European cities that ran from 2006–2008 and sought to integrate sustainability principles into core city decision-making processes. MUE-25 was followed up by CHAMP, Local Response to Climate Change, which seeks to integrate cities' management response to climate change, build capacity through national training hubs (including one in Germany), and further the EU's climate policy. Both are linked to the EU's 6th Environmental Action Plan and its Thematic Strategy on the Urban Environment.

In addition to capacity-building, the European Union has several programs that fund sustainable urban development. JESSICA (Joint European Support for Sustainable Investment in City Areas) is a European funding mechanism for sustainability projects in urban areas, providing loans for projects such as brownfield redevelopment, energy efficiency improvements, and transportation, water, and energy infrastructure. The Commission's has dedicated funding for research related to "smart cities," sustainability-related issues, and urban sustainability (Zimmermann, 2007). Clever cities can use these resources to fund local sustainability initiatives.

This list is by no means exhaustive, but highlights some of the mechanisms by which the European Union impacts sustainability in German cities. This brief overview highlights the significantly larger pool of financial, ideological, and institutional support provided to German cities from higher levels of government than is the case in the United States. In addition, it highlights the direct role the European Union plays in advancing sustainability in cities, bypassing the German national government in many cases.

Summary: Sustainability Governance in Germany Today

Germany became interested in sustainability at the national level later than the United States, but has since created several key institutions, policies, and funding mechanisms to promote its adoption and principles. In addition, the European Union has several tools to stimulate adoption of sustainability in European cities, many of which have been taken advantage of in German cities. Although it appears that some national and EU-level policies have not necessarily trickled down to the city level, it is clear that German cities have more support from higher levels of government—in terms of information, institutional support, and funding—than U.S. cities.

Despite criticism of German and European approaches, real social and political change towards sustainability is evident in Germany, particularly when compared to the United States. This can be seen in sustainability strategies at both the German and European level. The European Union adds momentum to German action and provides resources and support directly to cities, also its continued support of economic growth policies has been challenged as counterproductive to sustainability. Several additional factors also contribute to sustainability in Germany: its population density, transportation network, and social policies make the transition to sustainability less controversial than in the United States.

Avenues for Comparison: Studying Urban Sustainability in the United States and Germany

This chapter presented key background information about the United States and Germany. It highlighted several similarities between the two countries, but also some important differences that impact how cities are able to incorporate sustainability principles into political decision-making. Both countries are high-consuming federal democracies. But Germany's higher density and older development have led to a built environment that includes more comprehensive transportation system (in contrast to the car-dependent United States) and powerful, mid-sized cities with mixed-use neighborhoods. At the same time, interest in sustainability has diverged in recent decades. Whereas Germany has begun to institutionalize sustainability through its national sustainability strategy and a number of advisory councils, brief attempts at national attention in the United States were ended by indifference or even opposition. Finally, German national urban policy, with its planning mandates and normative pursuit of compact urban form, is more supportive of urban sustainability than the United States, where the federal government has abdicated its potential planning role and instead uses a "stealth urban policy" that supports suburban development at the expense of cities.

However, national-level differences play a smaller role at the city level than one might expect. Although German cities enjoy more support from the federal government and the European Union,[6] U.S. cities are also engaged in sustainability issues. The main challenges faced by cities in both countries are similar, as are their approaches, which focus on policy areas such as urban planning, climate mitigation, and economic development. In general, despite diverging supralocal political contexts, cities in both countries face similar challenges and are engaging in similar efforts to transition towards sustainability, making them worthy of further study and a comparison potentially quite fruitful.

6 Schreurs, Selin, and VanDeveer (2009a) highlight the engagement of multiple agencies and institutions related to sustainability in the European Union as compared to the United States, where interest was not only short-term but restricted to the executive branch (28). The same point could also be made in comparing the United States to Germany.

PART II
Urban Sustainability Governance in Six U.S. and German Cities

Chapter 4
The Instruments of Urban Sustainability Governance

As stated in the introduction, humanity currently faces a number of crises: a complex ecological crisis that includes biodiversity loss, multiple forms of pollution, and climate change; the continuing effects of a financial crisis that affect the poor and working classes in industrialized nations; and growing unrest in the Global South over economic and social policies that have not reduced poverty or alleviated inequality with the Global North. Sustainability can address these problems by changing how they are understood (Pettibone, 2014b) and includes five broad principles presented in Chapter 2. This has led many advocates to argue that a transition to sustainability is urgently required (for example, Jackson, 2009; Meadows, Randers, and Meadows, 2004; WBGU, 2011). But what would such a transition look like? What sorts of effort are required? And how could, for example, cities' progress toward this transition be measured, evaluated, and understood?

The answers to these questions are a source of academic debate. On one hand, proponents argue for the need for transition, leading to emerging literatures on sustainability transformation (McCormick et al., 2013; WBGU, 2011) and sustainability governance (Adger and Jordan, 2009; Lafferty, 2004; Voß, Bauknecht, and Kemp, 2006). On the other, much writing on sustainability governance focuses primarily on development of the right policy instruments, without examining how they are implemented or whether they are sufficient to ensure a successful transition (Bressers, 2004; Saha, 2009). Governance tools such as strategic plans, sustainability indicators, and new intergovernmental or interagency institutions, as well as economic instruments such as subsidies, tax incentives, and emissions or use caps, are seen as key ways to effect a sustainability transition. At the same time, growing empirical evidence has shown that these instruments often fall short of their perceived capacity to drive change. This raises the question: if governance instruments are not sufficient for the transition to (urban) sustainability governance, what is needed?

This question forms the backbone of this research project. In this chapter I discuss the research literature on (urban) sustainability governance. Despite calling for "a great transformation" described as encompassing all aspects of social and political systems (WBGU, 2011), the literature focuses almost exclusively on policy mechanisms implemented by governments and the role of government actors. However, empirical studies on the role of strategic plans and urban sustainability indicators, two governance instruments seen as important in urban sustainability governance, finds only limited efficacy, particularly in cases with low political interest in sustainability. Plans and indicators alone cannot guarantee a sustainability transition. Key actors must also be motivated to achieve urban sustainability governance.

What is (Urban) Sustainability Governance?

The importance of a transition to sustainability has led to a growing body of research on what such a transition would look like and how it could be achieved. The social and political change required has been conceptualized in several ways: sustainability transformation, socio-technical transitions, urban sustainability, reflexive or sustainability governance, and more (for example, Adger and Jordan, 2009; Brand and Fürst, 2002; McCormick et al., 2013; Voß, Bauknecht, and Kemp, 2006; WBGU, 2011). To some extent, these are different ways of saying the same thing, but the different terms also have subtle differences in emphasis, such as between technological, social or cultural, and political processes.

Sustainability governance falls within the larger umbrella of sustainable transition or transformation. McCormick et al. (2013) introduce a special issue of the *Journal of Cleaner Production* on sustainable urban transformation by highlighting "governance and planning" as one of four key drivers of such transition. To them, radical change requires not only technological solutions, but social, organizational, cultural, and political change as well (5). The German Advisory Council on Global Change's (WBGU, 2011) recent report on transition calls for "a new social contract for sustainability." The report sees a need for dramatic systemic change that affects politics, economy, and society, in order to reach a low-carbon future. It cites Karl Polanyi's description of the industrial revolution in calling the transition to sustainability a second "Great Transformation" (WBGU, 2011, 5). The transition to sustainability is thus all-encompassing: it affects every aspect of human life and system.

So a transition to sustainability calls for dramatic change in social, economic, technological, and political systems. Governance is thus an important site of transformation. Sustainability governance does not focus on technological aspects of transformation or on the mechanics of economic transition. Instead, the literature on sustainability governance—and particularly urban sustainability governance—focuses on the process through which sustainability principles are incorporated into political decision-making and social and cultural processes related to governance (Adger and Jordan, 2009; Lafferty, 2004; Voß, Bauknecht, and Kemp, 2006). The emphasis here is on the transition required in governance processes as part of a larger socio-political transition to sustainability. Sustainability governance includes instruments such as strategic plans, integrative policies and institutions, sustainability indicators, as well as broader institutional change and relational change between different actors.

Understanding Governance in the Service of Sustainability

In order to understand sustainability governance, it is important to first understand the concept of governance. A normative concept like sustainability, governance is about interactions between actors seeking to solve social problems. It includes multiple levels, from the instruments used to institutional arrangements and rules to the goals of governance itself.

Jordan (2008) argues that governance has enjoyed a similar history to sustainability: both are explicitly normative and vague concepts that have resonated across scientific disciplines and become popular since the 1980s. Kooiman (2003) describes governance as fundamentally about interactions:

> Governing can be considered as the totality of interactions, in which public as well as private actors participate, aimed at solving societal problems or creating societal opportunities; attending to the institutions as contexts for these governing interactions; and establishing a normative foundation for all those activities.
>
> Governance can be seen as the totality of theoretical conceptions on governing. (4)

Kooiman's understanding of governance—although admittedly more a collection of concepts than a framework for analysis—is often cited in sustainability governance research (for example, Gehrlein, 2004; Schindler, 2011), and serves as an important conceptual foundation. What Kooiman achieves is a synthesis of diverse literatures into a comprehensive conceptual framework. The governance context is marked by diversity, dynamics, and complexity. Interactions mutually influence elements (or instruments and tools), modes (self, co-, and hierarchical), and orders of governance.

Governance is a broad concept, including not only political institutions, but also market and civil society actors who seek to affect the images and instruments of governance. Kooiman's framework includes three orders of governance. The first focuses on the images and instruments: the tools used in interactions. Images are used to construct governance problems in a particular way. Problems with strong moral overtones can be especially difficult to resolve, because moral diversity is guaranteed and opinions are based on principles rather than interests (Kooiman, 2003, 145). Instruments are the tools governors use to affect social systems, and are a particular focus of the sustainability governance literature. The second order of governance, institutions, affects the ways in which actors can interact, as well as determine their needs and capacities.

Kooiman's third-order or meta governance includes the normative goals of "governors": in other words, the rules and goals of governance itself. Authors such as Gehrlein (2004) and Schindler (2011) have argued that sustainability governance includes new meta-level goals for governance. It is at this level that the rules of governance are determined, affecting the instruments and institutions available to governors.

Jordan's (2008) review of the literatures on governance and sustainability finds that governance, like sustainability, is a tricky concept: popular because it bridges disciplines and scales and thus imprecise. He discusses an emerging normative literature focused on identifying principles of "good governance" that promote desirable outcomes (Jordan, 2008, 23). This could also be understood as linked to Kooiman's (2003) third-order or "meta governance," as well as to sustainability governance.

Sustainability governance requires integration of disparate policy sectors and actor groups towards normative goals and as such functions at a meta level above policymaking. This makes it a natural fit with the concept of governance, itself an overarching strategy that incorporates explicit government actors as well as civil society and other stakeholders engaged in the more general business of governing (Jordan, 2008; Kooiman, 2003). In addition, the UN framework argues that good governance is an important precondition for urban sustainability (Luhde-Thompson, 2004, 482; UNCED, 1992). Following Kooiman's (2003) three orders of governance, sustainability governance focuses on three issues: new policies and instruments that promote incorporation of sustainability principles into decision-making; changes to organizational, legal, and societal institutions in order to promote such decision-making; and, related to the rule-making third order of governance, the principles of sustainability itself.

Grabow et al. (2011), in a report released by the German mayors' dialogue on sustainable development created by the German sustainability advisory council, emphasize the importance of understanding sustainability as a strategic process that coordinates political action. The report includes a graphic with arrows representing individual actors' directions, at first chaotic, but subsequently coordinated by sustainability (7). The authors call for sustainability to be elevated to a "*Chefsache*" (a top priority, or an issue for the "boss"). However, the fact that sustainability governance is a strategic process has made it exceedingly difficult to implement politically and study empirically. Lafferty and Meadowcroft (2001) highlight three challenges for sustainability governance:

> Any assessment of the future of sustainable development must take into consideration three dominant features of the idea: that it involves (1) goals and values which are at once *normative* and *relatively vague* with respect to specific policy prescriptions; (2) a 'programme' which has arisen *external* to normal national policy arenas; and (3) a political commitment which—largely due to the first two characteristics—is relatively 'soft' as a mandate for change. (456, emphasis in original)

In other words, sustainability is a normative appeal that has come from outside normal political channels and calls for the wholesale transformation of political institutions toward goals that remain unclear. Even as researchers and advocates see sustainability as something that must be integrated into the very fabric of governance, it has come from outside traditional governance institutions and therefore must somehow be incorporated into them as well (see also Lafferty, 2004). This has complicated the process of institutional change seen as necessary by advocates.

Governance is the study of interactions between actors seeking socio-political change. It includes a rich array of concepts of how these interactions take place, from image formation and development of governance instruments, to the institutional setting and its impacts on how actors interact, to the normative rules and goals of governance itself. Sustainability governance can thus be understood as linking the governance framework to the explicit goal of a transition to sustainability.

Mapping Urban Sustainability Governance

Urban sustainability governance is a more specific body of literature focusing on how sustainability governance is realized at the city level. It generally supports the perspective of the more general writing on sustainability governance: that paradigmatic change is required that includes new instruments and institutions and efforts inside and outside government. Much work digs deeper than more general sustainability governance work, however, to use empirical evidence to discuss the extent to which cities are incorporating sustainability into governance processes.

The literature focuses particularly on three areas. First, more theoretical work seeks to build an analytical framework for study. Evans et al. (2006) create a typology that focuses on a city's institutional (governmental) and social (for example, civil society, business, media) capacities. These capacities include committed individuals; alliances between actors or institutions; supportive media, business, or educational sectors; and civil society groups. Smith and Wiek's (2012) model of urban sustainability governance begins with sustainability principles, which are used to develop a vision that includes targets, implemented through good governance principles (social learning, consensual decision-making, adaptive capacity, openness, and accountability) with the help of local institutions.

Much research in urban sustainability governance relies on case studies. The second area of focus analyzes the effect of examining city characteristics such as size, institutional capacity and support, and political culture on sustainability (for example, Garcia-Sanchez and Prado-Lorenzo, 2008; Jepson, 2004; Newman and Dale, 2005; Portney, 2003). McCormick et al. (2013) see sustainable urban transformation, which includes governance, as having two dimensions—drivers of radical change (that is, governance and planning, innovation and competitiveness, and lifestyle and consumption) and urban structures (that is, resource management and climate mitigation/adaptation, transport and built environment). To them, cities face inherent challenges of "unsustainability" that demand new modes of governance that draw more on strategic, visionary planning, transition management, disruptive activities, and entrepreneurial change agents (5). Key factors are collaboration that builds learning, multi-shakeholder partnerships;

better communication between researchers, local communities, and governments; and more rigorous ways to evaluate sustainable urban transformation based on key principles and indicators. In short, what is needed are new visions, as well as more strategic interactions between actors, effective instruments, and the engagement of diverse actors in and outside of government to realize these visions.

Finally, a third area focuses specifically on one factor seen as important in the first two: strategic governance instruments. Such instruments can create an overarching vision for urban decision-making (for example, Innes and Booher, 2000; Pastille, 2002; Portney, 2003; Wheeler, 2004). Because sustainability is not a single policy area, but a new way of linking diverse problems, such instruments are inherently cross-cutting. They should connect sustainability to local policy problems, promote collaboration between societal actors and policy sectors, and facilitate strategic decision-making in the service of sustainability goals. Various instruments include strategic assessments and multi-criteria analysis, plans, interagency working groups or high-level sustainability offices, regulatory and economic instruments, and sustainability indicators. Two instruments in particular have been seen to play a key role in this process: strategic plans and sustainability indicators. Both are seen as tools at a meta or strategic level that aim to integrate policies and develop a more holistic approach to governance. Increasingly, U.S. and German cities have begun to develop and use these instruments.

How do these instruments contribute to urban sustainability governance? Strategic plans (comprehensive plans in the United States) have gained prominence in recent years both as good planning practice and an instrument of urban sustainability governance. Unlike narrower land-use or zoning plans (*Flächennutzungs-* and *Bebauungspläne* in German), strategic plans work at the meta level of city planning, serving as a medium- to long-term vision statement and including policy areas beyond the built environment. Such strategic visions go by many names—including sustainability plans, vision documents, and urban development plans—but because of their similar functions will be considered equivalent to strategic plans here.[1] Their role is seen as elaborating a city's vision of sustainability at the strategic level, which can then be implemented through various policy mechanisms.

Sustainability indicators have also been lauded as an important tool of sustainability governance. Researchers have listed a number of purposes: they make the vague idea of sustainability tangible; they allow researchers and scientists to play an important role in information creation and policymaking; and they seek to reconcile the tension between a need for clear definition and the contested nature of sustainability (for example, Bell and Morse, 2001; Gehrlein, 2004; Innes, 1990). They thus serve as a translating mechanism by operationalizing a

---

1 The six case cities studied used different names, such as comprehensive plan, growth management plan, vision (*Leitbild*), and Local Agenda 21. In presenting the cases, I will generally refer to cities' plans using their own terminology, but use "strategic plans" when speaking generally about such plans.

normative vision of sustainability with the help of empirical data. This role, as well as their versatility, makes sustainability indicators an important instrument of urban sustainability governance.

In both the United States and Germany, researchers have had high hopes that strategic plans and sustainability indicators can effect a sustainability transition in cities (Beatley, 1995; Feitelson, 2004; Gahin, Veleva, and Hart, 2003; Gehrlein, 2004; Meadows, 1998). But both instruments have to some extent failed to reach their potential use (Gehrlein, 2004; Gibbs and Krueger, 2007; Pastille, 2002). Although much empirical literature exists that attempts to explain the role these instruments play in urban sustainability governance and what factors affect their implementation—from inclusion of key stakeholders to political leadership to city characteristics such as education, major industries, and urban form—findings to date suggest that sustainability is not necessarily the main goal of cities using these instruments (for example, Innes and Booher, 2000; Gibbs and Krueger, 2007). In addition, numerous cities have seen implementation problems with both plans and indicators (see Chapter 2). Thus, their use is subject to questions of implementation and evaluation. First, what factors impact the effectiveness of plans and indicators? Second, even if these instruments are being used, how are they contributing to urban sustainability governance?

Urban sustainability governance calls for normative change in social and political interactions by diverse actors toward the goal of a sustainability transition. It is to some extent vague and has come from outside traditional political channels. It requires a strategic outlook and new institutions and instruments to integrate disparate policy sectors, connect various actors, and coordinate comprehensive action. The next section goes into more detail on one aspect of this complex change process: instruments of urban sustainability governance, specifically, strategic planning and sustainability indicators.

The Meta-level Instruments of Urban Sustainability Governance

The last section discussed the strategic nature of urban sustainability governance as well as its interest in more inclusive decision-making processes. This section presents two instruments—strategic plans and sustainability indicators—that are considered important in promoting urban sustainability governance. Drawing primarily from experiences in the United States and Germany, it discusses both potential roles of plans and indicators as well as empirical findings that show how these instruments have fallen short of expectations. This raises the question: if governance instruments alone are insufficient, what is needed? Two factors are highlighted as important for further examination: conflicting visions of sustainability and lack of interest in implementation.

Strategic Plans

Planning is noted by many as a useful way to promote sustainability in cities, for several reasons. First, planning is long-term and visionary, making decisions based on decision-makers' vision for the city's future (Feitelson, 2004; Wheeler, 2004). Planning is also holistic, bringing together policies from multiple city agencies and often including the public (Beatley, 1995; Innes and Booher, 2000; Moss and Grunkemeyer, 2010). Higher levels of government can also require comprehensive land-use plans (Larsen, 1996; Miller, 2004), making them a way to incorporate sustainability into decision-making related to the city's physical form. Such requirements can also include strategic elements. Finally, best practice principles of urban planning—particularly used by advocates of new urbanism, smart growth, and *Baukultur or "building culture"*—generally fit well with policy approaches that are considered to promote urban sustainability (Godschalk and Anderson, 2012; Selle, 2006). These features make strategic plans a useful tool to incorporate sustainability principles into political decision-making processes at the city level. This section looks specifically at the role of plans in promoting urban sustainability. It examines research on strategic, comprehensive, and sustainability plans as well as Local Agenda 21s.

Planning and sustainability formed an early, positive relationship in the United States. Godschalk and Anderson (2012) argue in the final report of the American Planning Association's (APA) Sustaining Places Task Force that planning can play a special role in implementing sustainability at the urban level, because planning "has the advantage of being able to integrate long- and short-range perspectives and to coordinate other policies, plans, and programs into a single accessible document" (8). Their report follows a recent trend in several planning movements, such as smart growth and new urbanism, which have sought to challenge the sprawling suburban development that has characterized U.S. land-use change in the past half-century (Wheeler, 2004). These planning movements, although interested in human-scale urban form for aesthetic and quality-of-life reasons, call for policies such as integrated transportation planning, walkable neighborhoods, and mixed-use development that have also been taken up by urban sustainability advocates (Jepson and Edwards, 2010; Kooshian and Winkelman, 2011). In this way, planning is used to shape the urban environment to be more conducive to sustainability.

Germany has seen a similar rise in interest in strategic planning, despite a different urban context and a stronger legal planning framework. Federal law requires cities to develop comprehensive land-use (*Flächennutzungspläne*) and zoning (*Bebauungspläne*) plans linked to state and regional plans (Light, 1999; Scholl, Elgendy, and Nollert, 2007; see also Chapter 3). These plans have historically required cities to develop in certain ways and ensure "equivalent living conditions" through regional coherence in terms of, for example, transportation and housing, and involve the more active involvement of multiple levels of government (Larsen, 1996).

In the last two decades, the vision behind such operational plans is being aligned more explicitly to sustainability. The relatively young *Baukultur* (roughly, "building culture") movement has sought to bring order to architecture and urban planning by developing principles similar to movements in the United States (Hackenberg, Oostendorp, and Wiegandt, 2010). The *Bundesstiftung Baukultur*, a foundation created to disseminate the movement's principles, calls for "*Nachhaltigkeit und Dauerhaftigkeit*" (sustainability and longevity) and seeks to find harmony between environmental, social, and economic goals (Bundesstiftung Baukultur, 2014). Sustainability is also mentioned in German federal planning law, which calls for economic, social, and environmental needs to be harmonized (Bückmann, 2002, 156). The final report from the German sustainability advisory council's dialogue of mayors argues that in addition to such legal planning requirements, strategic plans are an important tool in realizing urban sustainability (Grabow et al., 2011).

In short, strategic plans are considered a key way to incorporate sustainability principles into urban decision-making processes, and have been advanced in both the United States and Germany in recent years. These countries come from different planning contexts: German cities are required to develop and regularly update comprehensive land-use and zoning plans, whereas U.S. planning requirements vary between the states and often include only zoning designations. Still, cities in both have begun to use strategic planning to advance sustainability, whether by shaping the built environment or developing a more general political vision.

However, empirical research on how plans actually contribute to urban sustainability governance has shown mixed results. Berke and Conroy (2000), in an analysis of 30 strategic plans in U.S. cities, found no significant difference between cities that used the term "sustainability" and those that did not: Jacksonville, Florida, which uses a quality-of-life framework, received the highest score on their metrics.[2] Several years later, however, the same authors found that an explicit focus on sustainability in planning documents aided the sustainability of the eventual plan (Conroy and Berke, 2004, 1384–5). They also found that state comprehensive planning mandates can substitute local capacities or broad participation, and that sustainability principles do not seem to be fully understood by planners.

Staley (2006) is more critical, finding in a study of sustainability planning in Santa Monica that the city's sustainability framework is nothing more than an

---

2 However, the authors' approach highlights a major problem with defining urban sustainability in research projects. They use six principles of sustainability based on the notion of the city as "reproducer": harmony with nature, livable built environments, place-based economy, equity, polluters pay, and responsible regionalism. These principles are skewed toward the environment and differ from those recommended by both U.S. and German planning and sustainability organizations. Still, this demonstrates the importance of their finding: if a city using a quality of life framework does better with ecological measures, sustainability may not add much value to planning.

"aggressive environmental policy agenda" (114) that does "not necessarily promote sustainable development" (113). Gibbs and Krueger's (2007) study of Austin and Boston, by contrast, finds that sustainability often takes a backseat to economic development goals, arguing that the strength of neoliberal growth coalitions drowns out sustainability concerns. This research highlights the malleability of the sustainability concept within the literature and the political sphere.

Although sustainability has proven more popular in Germany with the diffusion of Local Agenda 21s and other tools, research on cities there has shown problems with implementation. Beuermann (1998), in an early study of Local Agenda 21s in German municipalities, highlights three main impediments to implementation: lack of information on LA21 and how to implement it; interest in other priorities, such as public budgets and economic development; administrative structures that inhibit successful policy integration; and insufficient support by local actors. Hackenberg, Oostendorp, and Wiegandt (2010) find only spotty implementation in their examination of strategic plans in 20 major German cities. The authors link success to connection between levels of planning: specifically, vision documents must be tied to legally mandated land-use plans in order to have maximum effect (echoing Conroy and Berke, 2004, above).

Strategic plans in Germany are voluntary instruments that run parallel to the comprehensive, land-use, and neighborhood plans required by federal and state law (Larsen, 1996; Scholl, Elgendy, and Nollert, 2007). Altrock (2004) cautions that vision documents are often overvalued and can prove ineffective when they try to serve both analytic and normative purposes, for example, when restricted to expert planning goals but still trying to build consensus among local actors. Perhaps because German cities are already legally required to develop expert planning documents, it is important for them to clearly identify a unique role for sustainability plans.

Finally, although Local Agenda 21 has provided a framework for action in Germany, it appears to have run its course. Kern, Koll, and Schophaus (2007) find that the majority of initiatives were developed between 1997 and 2002, with interest dropping off afterwards. An administrator involved with sustainability in Hamburg told me that, although "Local Agenda 21 is out," sustainability is still very important in the planning community (HH01). In the United States, a lack of national resonance for sustainability issues has not stopped the trend toward strategic and sustainability planning at the local level (Godschalk and Anderson, 2012; Portney, 2003; Wheeler, 2004). However, the changing popularity of sustainability and Local Agenda 21 point out the importance of political relevance to local actors.

The findings related to strategic planning for sustainability in the U.S. and Germany point to related phenomena. Specifically, although strategic plans may be able to aid the incorporation of sustainability principles into urban decision-making, they do not appear to completely fulfill this role. Problems appear in two key areas. First, strategic plan goals may bear little resemblance to sustainability principles. Second, even if the plan does promote sustainability, lack of interest or understanding can inhibit its successful development or implementation. The trend of developing Local Agenda 21s, which never caught on in the United States,[3] appears to have faded in Germany. This lack of efficacy is echoed in research on sustainability indicators, as seen in the next section.

Sustainability Indicators and Urban Governance

Sustainability indicators[4] have been a major focus of sustainability research, examining issues including 1) development of indicators and indices to measure complex ecological and social phenomena; 2) discussion of frameworks or purposes of indicators; and 3) empirical study of how indicators affect policymaking (see for example, Holman, 2009 for an excellent review). This research is interested specifically in how city-level sustainability indicator sets contribute to urban sustainability governance. Although sustainability indicator sets vary dramatically in size and scope, intended use, and developer, they are generally a group of quantifiable measures that seek to operationalize sustainability in the urban context in a way useful for decision-making.

The literature on sustainability indicators began with ebullience regarding their importance with respect to sustainability transition. Meadows (1998), in an early report on the importance of indicators, proclaims:

> Indicators don't guarantee results. But results are impossible without proper indicators. And proper indicators, in themselves, can produce results. (xii)

Although this can be considered an especially optimistic view of indicators' value, it highlights the key role that indicators are seen to play in sustainability governance. Fürst (2002) sees target indicators—sustainability indicators that include a target or goal—as an important way to support the "paradigmatic governance" of sustainability (see also Kopfmüller et al., 2001). Such indicators translate strategic goals into specific targets for action, increasing the monitoring potential of the system (Fürst, 2002, 184f). However, coming to the necessary

3 U.S. cities' interest in sustainability has taken other forms, such as strategic or sustainability plans and membership in organizations such as ICLEI and the U.S. Mayors' Climate Protection Agreement.

4 This overview focuses almost exclusively on the use of sustainability indicators in local policymaking processes. The literature is in fact many times larger than this, and includes numerous other issues not discussed here, such as methodological considerations related to construction of individual indicators, indicator sets, and indices (Bossel, 1999, 2001; Diefenbacher and Zieschank, 2011; Magee and Scerri, 2012; Mori and Christodoulou, 2011); indicator use to rank cities' sustainability (Dovern, Rickels, and Quaas, 2012; Häberli and Schneider, 2002; Siemens, 2011a); and indicator-related issues such as integrated impact or sustainability assessment (LIAISE, 2010).

agreement over targets requires not only the process for goal development but also a desire for strategic management in the first place (Fürst, 2002, 185).

Gehrlein (2004) elaborates on this in his synthesis of indicators' key governance functions.[5] According to his review of the literature, sustainability indicators can:

- describe and report on the current state of the city/country/etc.;
- forecast future problems;
- point out areas for action;
- communicate and build awareness on sustainability-related issues;
- aid political and administrative decision-making;
- formulate and concretize sustainability goals;
- manage performance; and
- serve as benchmarks for comparison with others (that is, other cities). (from Table 2.4 on p. 43; my translation).

Indicators are thus a valuable tool to incorporate sustainability into political decision-making processes (Rydin, Holman, and Wolff, 2003). They can help set the political agenda (Pinfield, 1997), draw comparisons (Pastille, 2002), and set a strategic course for policymaking (Schindler, 2008; Teichert, 2000).[6] In addition, researchers have suggested more strategic purposes for indicators, such as integrating the work of different administrative fields in local government (Gehrlein, 2004), communicating with the public (Pinfield, 1997), and building consensus through participatory processes (Innes and Booher, 2000; Köckler, 2005; Parris and Kates, 2003).

Sustainability indicators are also seen as able to nail down a definition of sustainability useful to policymakers (Feller-Länzlinger et al., 2010; Gehrlein, 2004).[7] Journel et al. (2003) refer to sustainability indicators as "hybrid objects":

---

5 Gehrlein's review is more concentrated on the German-language literature, which deals with many of the same concerns as the Anglo-American literature on sustainability indicators. However, some differences in emphasis can be noted, such as a stronger inclination toward participatory processes in the U.S. literature, matched by a larger focus on indicators' role in public administration in Germany, as seen by Gehrlein's list of governance functions. This difference in emphasis is also reflected by the use of indicators in my six case studies, discussed in Chapter 6.

6 Indicators may also be desirable because they provide a clear role for scientific experts to assist in decision-making: numerous academics have put forward both general frameworks for indicators or indices (for example, Bossel, 1999, 2001; Becker, 2005; Mori and Christodoulou, 2011) and proposed indicator sets for specific cities (for example, Block, Van Assche, and Goeminne, 2013; Diefenbacher et al., 1997; Häberli and Schneider, 2002).

7 The assumption that a lack of objective technical information is the source of bad policy, as well as the confidence that more scientific findings will fundamentally change political decisions, has been challenged numerous times in the past several decades (for example, Fischer, 2009; Innes, 1990; Kahan et al., 2013; Lindblom and Cohen, 1979; Weiss, 1979).

they seek to reconcile the contested concept of sustainability with a scientific interest in objective data. Bell and Morse (2001) describe this as a "conundrum":

> Herein we believe rests the central conundrum of SIs [sustainability indicators]. We are trying to use SIs as a tool to gauge something that is highly subjective and ridden with human values and desires, yet we are trying to do this either by ignoring these very human aspects or by trying to reduce them to a few simple numbers (Bell and Morse, 2001, 307).

In other words, sustainability indicators try to square the circle: they seek to create an objective understanding of the contested nature of sustainability in order to create tools compatible with governance and policymaking. They can thus be understood as a translating mechanism, a bridge between the constructed nature of sustainability and the need to reach definitional agreement in order to make policy decisions.

Empirical research on sustainability indicators has seen a lack of implementation similar to that of strategic plans presented above.[8] Numerous case studies have shown that indicators are not used by policymakers (Holman, 2009; Misaras, 2007; Rosenström, 2006). Others express disappointment that indicators have not changed attitudes and behavior (Yli-Vlikari, 2009) or fulfilled their potential integrating function within government structures (Gehrlein, 2004). However, the studies that saw indicators as the least effective were also the most restrictive in their understanding of indicator use. Such research generally understood indicator use as effective only when indicators were directly linked to policymaking. The main problem, according to these studies, is indicators' multiple incompatible functions, which prevent them from being meaningfully communicated between actors (Yli-Vlikari, 2009, 899).

Researchers' disappointment may be more appropriately directed to political decision-making in general rather than sustainability indicators. Some scholars have argued that information rarely sees a direct use in policymaking (Innes, 1990; Lindblom and Cohen, 1979; Weiss, 1979). Instead, decision-makers select and process information along two dimensions—truth and utility—based on both its perceived accuracy and relevance to decision-making (Weiss and Bucuvalas, 1980). In addition, there is a strong values component to information use: values influence how new information is processed, and information in turn influences values, albeit in ways that are difficult to observe (Kahan et al., 2013; Lindblom, 1990). Kahan et al. (2013) recently showed how partisans misinterpret statistical problems to support their political views, highlighting the interpretive nature of even supposedly neutral data. At the same time, information can serve an "enlightenment" function, subtly altering one's worldview over time (Weiss, 1979; Weiss and Bucuvalas,

8 This disappointment also echoes that of social indicators (see: Cobb and Rixford, 1998; Innes, 1990) and implies the inherent problem of incorporating indicators based on essentially contested issues into policymaking.

1980). Weiss (1979) points out seven models of how social science knowledge is used in decision-making, including direct use, to support political goals, and learning functions such as in groups. Her "enlightenment" category refers to general social science "concepts and theoretical perspectives … [that] permeate the policy-making process" (429); here, knowledge changes assumptions, analytical models, and worldviews in a nearly invisible process over years or decades. Indeed, increased acceptance of sustainability principles at all could be considered related to this enlightenment function.

Studies that use a broader definition such as that of Weiss have highlighted important effects in public awareness and issue framing (for example, Bell, Eason, and Frederiksen, 2011; Holden, 2007), as well as networking and coordination between stakeholder groups (for example, Innes and Booher, 2000). For example, Rosenström (2006) analyzes how Finnish environmental policymakers use sustainability indicator reports to make decisions. She finds that politicians often use indicators in order to make themselves more credible, indicating a primarily symbolic use. She cites Weiss's (1983) argument that ideology, interests, and information are the three driving forces of decision-making (Rosenström, 2006, 10). Finally, Rosenström argues that "indicator providers" have an important role to play in selecting politically relevant indicators, updating them in a timely way, and disseminating them to potentially interested parties. This supports early assertions of Weiss (1979) and Lindblom (1990) that political beliefs affect how we process information, which suggests a more complex reality for use of sustainability indicators. Indeed, as seen with comprehensive plans earlier, it appears that values play an important role in the use of sustainability indicators in decision-making.

Cities in both the United States and Germany have been leaders in developing sustainability indicators. Seattle developed perhaps the world's first sustainability indicator set in 1993 (Holden, 2007), which served as a model for efforts in Germany (Lang, 2003). Case studies in both countries reflect the diversity of indicator efforts both in functions and involved actor groups, although effects of the latter have not been an explicit focus of the literature. Indicator researchers have demonstrated primarily intangible effects (Gahin, Veleva, and Hart, 2003; Innes and Booher, 2000; Köckler, 2005).

The literature on urban sustainability indicators discussed here highlights both the hope of many scholars that indicators can help incorporate sustainability principles into decision-making, as well as the uneven experience cities have had using such indicators as a tool of urban sustainability governance. Indicators are "hybrid objects" that seek to translate the vague concept of sustainability into empirically verifiable quantities. As such, sustainability indicators serve multiple different incompatible purposes, and rarely are used directly in policymaking, confounding researchers' expectations. Additionally, research on the use of information in decision-making finds that policymakers most often use information to support their values and advocate for their political goals. Therefore, as with strategic plans, motivation appears to play an important role in ensuring the efficacy of sustainability indicators in urban sustainability governance.

Conclusions: Reconceptualizing Urban Sustainability Governance

Because sustainability calls for fundamental change to policymaking processes—images and instruments to address sustainability issues, new institutions that cut across traditional political sectors, and reorientation of governance goals towards sustainability—it has been argued to require new forms of governance (Brand and Fürst, 2002; UNCED, 1992) or a societal transition (WBGU, 2011). Sustainability governance deals with the process of incorporating sustainability principles into political decision-making. The literature has increased understanding of governance structures and highlighted several instruments in urban sustainability governance.

Two governance instruments—strategic plans and sustainability indicators—are seen as playing an important role in urban sustainability governance. Both tools can create a vision for policymaking at the strategic level that integrates diverse policy sectors as well as link the diffuse concept of sustainability to specific policy objectives. In this way, plans and indicators can translate the concept of sustainability into a concrete vision for policymaking. However, empirical researchers have been disappointed in how these instruments are actually used. Whether due to lack of use based on insufficient motivation of key actors or unclear connection to sustainability principles, neither plans nor indicators have fulfilled their potential as governance instruments.

This raises the question of why these instruments are developed and how they actually do contribute to urban sustainability governance. If their use fails to meet researchers' expectations, what is the purpose of plans and indicators? That is the question this study seeks to answer. The next section begins to do so by exploring the roles played by these instruments, as well as sustainability-minded groups, in six U.S. and German cities.

Chapter 5
Creating a Vision: Strategic Plans and Urban Sustainability

This book seeks to understand why cities use instruments of sustainability governance and how these tools incorporate sustainability principles into policymaking by examining the experiences of six cities in the United States and Germany. This chapter focuses on strategic plans, a popular instrument of urban sustainability governance. It seeks to address the following questions: *Why do cities develop strategic plans? How do plans help incorporate sustainability principles into political decision-making?*

As discussed earlier, sustainability provides a broad normative vision that is generally consistent with good planning principles (such as multi-modal transportation, compact cities, increased greenspace, and inclusive planning processes). This has provided planners with new language and goals to integrate into their work. Urban sustainability advocates also see strategic plans as an important tool to incorporate sustainability into urban decision-making.

Strategic plans are a tool of increasing importance in both the United States and Germany. They are used to develop a strategic vision for the city and can be a way to incorporate sustainability into urban planning (Godschalk and Anderson, 2012; Grabow et al., 2011). However, empirical studies on strategic plans reveal a more complex picture of their ability to contribute to urban sustainability governance (Gibbs and Krueger, 2007; Jepson, 2004). Studies in both the United States and Germany have found that development of plans may not be sufficient, as many have spotty implementation (Jepson, 2004; Kern, Koll, and Schophaus, 2007). Even if plans are implemented, it is not clear whether they contribute to urban sustainability.

This chapter explores these issues in more detail through the experience of six cities. It focuses on how and why plans were developed and how well they reflect sustainability principles. Each case describes the development of the plan, its main purpose, and the role of key actors involved in the process. Because not all cities studied saw the same level of input into or implementation of plans, the cases vary in length, but all focus on details of motivation and process.

Berlin: An Ambitious but Failed Local Agenda 21

Berlin was a flurry of sustainability activity in the 1990s and early 2000s, due to the massive political change spurred by city reunification but also an influx of funding dedicated to sustainability and LA21-related projects. For example, the Europan Commission supported development of the *BerlinStudie* (Berlin study) between 1998 and 2000. This strategic plan's two main goals and six principal elements demonstrated a clear interest in sustainability; the English term is even mentioned as a connecting thread (*BerlinStudie*, 2000, 4). These were divided into 12 points that focused on developing a thriving, innovative economy with good quality jobs that moves toward a "social and sustainable market economy"; building a socially just community that is strategic in welcoming immigrants and attractive to young people; reducing environmental and transportation stress; and improving political engagement with civil society, citizens, and the European community (*BerlinStudie*, 2000, 2–3). One unique feature is an understanding of Berlin as "both East and West," which leads to goals like increased internationalization, deeper cultural exchange, and improved international transportation links (*BerlinStudie*, 2000, 10).

The *BerlinStudie* called on the Berlin government to deliver a vision document (*Leitbild*) and a preliminary sustainability indicator set. These were developed through a participatory process led by the Agenda-Agentur Berlin and the Technical University and given to the city parliament in fall 2001 (Häberli and Schneider, 2002). This was the only report published on the plan. However, despite positive feedback from some parliamentarians, this study was not further used in decision-making.

Indeed, the failure of this externally developed plan foreshadows Berlin's experience with Local Agenda 21. The city's neighborhoods adopted LA21 early as a planning instrument. The first Local Agenda 21 (LA 21) in Germany was developed in the neighborhood of Berlin-Köpenick; all 23 of the city's neighborhoods developed LA21s between 1994 and 1998 (Beuermann, 1998; Kern, Koll, and Schophaus, 2007). The neighborhood LA21s were developed "bottom-up", using an "approach ... based on its strong tradition of citizen initiatives which pushed the Agenda 21 process from the outset" and funded largely by the federal government (Kern, Koll, and Schophaus, 2007, 612). A public working group at the city level was created in 1995 to promote cross-neighborhood exchange. Despite this, neighborhood efforts took a while to trickle up the city level. Efforts to develop a citywide LA21 began in 1997, when the urban planning department started a monthly roundtable on sustainable development in Berlin and Brandenburg; it then created an Agenda 21 office in 1999 (Abgeordnetenhaus Berlin, 2006; Kern, Koll, and Schophaus, 2007). The citywide effort was more top-down and professional, running generally independently of local neighborhood activities (Kern, Koll, and Schophaus, 2007). The process was also complex, and was driven by four groups: civil society and business organizations, the neighborhoods, the city administration, and the Berlin parliament (Abgeordnetenhaus Berlin, 2006, 4).

Kern, Koll, and Schophaus (2007) point to the city's "disastrous" financial situation as a major impediment to plan development and implementation (615). Between 1997 and 2000, the federal government provided about €600,000 in annual funding to coordinate activities in the neighborhoods. As soon as this funding was cut, “most of the activities broke down” (Kern, Koll, and Schophaus, 2007, 615). The city’s central Agenda office, housed in the Senate’s planning department, remained, albeit with a mere three and a half positions. Since 2000, coordination of the city’s LA21 has been delegated to external groups such as the Agenda Forum, the environmental group Grüne Liga, and Berlin 21. By 2014, all city funding for Berlin 21 had come to an end (BER03).

The Agenda office produced a first draft LA21 to the city’s urban development ministry in fall 2003, which was then reworked and submitted to the Senate in March 2004. The Senate gave a commented copy to the parliament, which was rejected. The planning committee subsequently edited the document into something that would get majority approval, the outcome of which is the current LA21. Much of the original draft was drastically changed, expanded or inserted (for example, new climate protection and economic sections), or removed (for example, sections on participation and the employment outlook) (Abgeordnetenhaus, 2006, 4–5).

The Berlin parliament finally passed the city’s Local Agenda 21 in June 2006, making it Berlin’s key strategy for local policy (Berlin Abgeordnetenhaus, 2006). In this decision, it determined to report every five years, beginning in 2009, on progress toward eventual 2030 goals (Berlin Senat, 2009).

> With the resolution on June 8, 2006, the Berlin parliament declared Local Agenda 21 Berlin the guiding principle for future city policy. The document includes a strategic purpose and concrete measures to implement Berlin’s sustainable development in an action program for the 21st century. LA 21 Berlin uses a comprehensive approach with the goal of integrating the foundation of sustainable development in as many policy areas as possible. The strategy includes guidelines and an action plan with seven thematic areas and two cross-cutting themes. (Berlin Senat, 2009, 3)

The report’s themes include social life in the city, public engagement and participation, transportation and mobility, Berlin in the region, energy and climate policy, economy and labor, and education, as well as the cross-cutting issues of gender equality and global responsibility (Berlin Senat, 2009, 7). However, political winds were not kind to the plan. The political coalition of SPD, Left, and Greens that got the city’s Agenda 21 passed dissipated when three key advocates left parliament soon after its approval (BER01, BER02). In addition, the plan was criticized for being “too detailed, not to the point” (BER02). One interview partner heard from a parliamentarian after LA21’s passage that he was glad the process was finally over, with the implication that the plan would need no further effort, such as implementation (BER04). Whatever the reason, a plan so many years in the making has seen only scattered relevance in agency offices committed to its realization (BER02).

In 2013, Berlin’s new urban development minister Michael Müller began developing a new urban development plan for the city entitled *Berlin 2030*. This plan seeks to serve as an orienting framework for political decision-making in the city connected to comprehensive planning documents, political and administrative actors, and the public (Senatsverwaltung für Stadtentwicklung und Umwelt [SSU], 2013, 5). Plan development was ongoing as of spring 2014 and included a series of public events and workshops. Public participation was limited, however to the "Berlintelegramm" (sticky notes on which members of the public could write their wishes for Berlin in the future), input from which was reduced in event documentation to either a speech picking out selected feedback (at the kick-off event) or a simple list of all comments (at the first city forum) (SSU, 2013; SSU, 2014b). The plan website summarizes the participative approach as follows:

> More than 2,500 Berliners took advantage of the opportunity to share their opinions and suggestions in the past 18 months and get involved in the discussion process by Berlin telegram, per Twitter and SMS. In addition, our internet presence provided access to extensive information on the current, ongoing development process. The 75,000 monthly visitors demonstrate the great interest of all participants. (SSU website, accessed March 6, 2015)

As will be seen in later cases, particularly in Heidelberg and Portland, this definition of participation is particularly narrow.

Critics worried during field research that plan implementation would be hampered by political problems—such as Müller's poor relationship with Mayor Wowereit (BER02)—an issue rendered moot by Müller's recent succession as mayor in December 2014 after Wowereit's resignation (Donahue, 2014). Development has also resulted in several sectoral plans related to business development, transportation, consumption and waste, housing, and urban centers. The city has also done ambitious work on urban biodiversity and climate change, working closely with leading experts at local universities in these areas (Reusswig et al., 2014; SSU, 2012).

The focus of *Berlin 2030* uses optimistic demographic projections to justify an active economic and urban development strategy (SSU, 2014). The first words of the urban development minister's foreword refer to growth:

> Ladies and gentleman, Berlin is a growing city. By 2030, 250,000 additional people will have moved to this exciting metropolis. We want—and need—to be prepared. (SSU, 2014, 1)

Berlin 2030 presents several strategies to make Berlin a "sustainable city," including promoting economic clusters for academic and artistic work, preserving neighborhood character, pursuing climate neutrality and integrated mobility, and engaging the public.

The draft plan as of spring 2014 appears to be more of a marketing instrument than a policy tool, which the initial document calling for plan development even alludes to: "The urban development plan *BerlinStrategie* is not a marketing instrument, but presents the foundational content for marketing activities" (SSU,

2013). As such, its vision is oriented beyond the city, raising questions about the extent to which, for example, population projections are to be seen as a given fact or a political goal. At the same time, the plan centers around eight strategies "for a sustainable city": three of which focus on economic growth (through scientific innovation, support for creative workers, and education) (SSU, 2014a).

The final plan, released in November 2014, is not connected to, nor does it acknowledge, the city's Local Agenda 21, which the new plan would be replacing (SSU, 2014). In addition, the final plan appears to vary little substantively from the initial draft presented in 2013, despite mentions of participatory process development (SSU, 2013; SSU, 2014). This suggests a disconnect between city decision-makers and Berlin residents, which is further supported by the Senate's surprise at the success of a public referendum overturning its plan to redevelop Tempelhof field, the site of a former airport that currently serves as a park.[1] People I spoke with linked this to politics: the new urban development minister sees *Berlin 2030* as a way to make a name for himself; key supporters of Local Agenda 21 have since retired or fallen out of favor (BER02, BER03).

Although Berlin's Local Agenda 21 was developed through a long-term deliberative process and closely linked to sustainability principles, it has not seen implementation. In addition, the city's new urban development minister (and now mayor) developed a new urban development plan that is not linked to the city's existing plan, in the form of its Local Agenda 21, and more tenuously connected to sustainability principles. The city's experience shows that high-level political support is more critical to plan implementation than an inclusive development process or explicit link to sustainability. These lessons echo the experiences of other cities studied here.

1 This is based on the author's experience at a meeting held by the Senate's SPD caucus after the referendum.

Table 5.1 Planning and sustainability in Berlin

Plan	Key features	Link to sustainability principles
Local Agenda 21 (2006)	• Development process included strong engagement of outside actors in working groups • Plan is comprehensive and seeks to guide all aspects of city policymaking	• *Long-term:* 25-year outlook • *Systems-thinking:* Cross-cutting themes acknowledge complexity • *Deliberative:* Achieved through thematic working groups and engagement of civil society actors
BerlinStrategie: Stadtentwicklungs-konzept Berlin 2030 (2014)	• Primarily focused on accommodating or appealing to new residents in targeted economic clusters • Strong engagement of outside experts, limited opportunities for public participation	• *Acknowledges limits:* Yes, through discussion of climate neutrality • *Long-term:* 15-year outlook • *Systems-thinking:* Acknowledges conflicts within and between goals • *Deliberative:* Limited public engagement; one thematic area calls for participatory planning • *Reflexive:* No acknowledgment of LA21

Hamburg: Developing Political Visions for Growth

In 1997, after signing the Aalborg Charter in 1996, the Hamburg Senate released its first major action explicitly related to sustainability and its only attempt at a Local Agenda 21, the *Action Program of Hamburg Agencies for Local Agenda 21* (*Aktionsprogramm Hamburger Behörden zur kommunalen Agenda 21 Hamburg*, Schindler, 2011). As its title suggests, this document focused on internal administrative actions and included 99 city government projects, most of which already existed—in some cases for years—that were linked in some way to Agenda 21 (Schindler, 2011). Menzel (2000), speaker of the *Zukunftsrat* (Future Council) a consortium of organizations created to coordinate sustainability in the city (see Chapter 7), was critical of the document, calling it "a bit weak as a definition for 'sustainable development'" (55). Rather than providing a strategic plan, the "action program" served essentially as a catalog of pre-existing measures.

This was followed in 2001 by the *Environmental Timetable: Goals for a Sustainable Hamburg* (*Kursbuch Umwelt: Ziele für ein zukunftsfähiges Hamburg*), an ambitious document developed by the city's urban development and environment agency (*Behörde für Stadtentwicklung und Umwelt*, or BSU) that was highly regarded by environmental NGOs and included concrete goals for ecological sustainability (BSU, 2001; Schindler, 2011). The document's genesis in the environmental agency meant that it had no power over other aspects of city policy, however. It aimed to support the overarching goals of local foresight and global responsibility, the necessity of which then–BSU minister Porschke explained with a sailing metaphor:

> So that the journey does not go off course, goals and interim targets must be determined .... If a captain doesn't know what shore he is aiming for, no wind is the right one (BSU, 2001, i).

The document's goals were linked to five main areas: ecological protection, resource conservation, climate protection, human health, and community quality of life. However, because it was commissioned under a Green minister, the document was considered dead after the 2001 elections put a coaliton formed by the conservative CDU, FDP, and Schill parties into power. Schindler (2011) suggests it may be "one of the most beloved politically dead documents in Hamburg" (181). Nevertheless, its origins in the environment ministry make it unclear how wide its effects could have been.

In 2002, the new conservative government related a new vision document entitled *Metropolis Hamburg – Growing City* (*Metropole Hamburg – Wachsende Stadt*), which seemed to come out of nowhere and in direct opposition to sustainability-related efforts by key civil society actors at the time (Schindler, 2011). Then-mayor Ole von Beust (2004) saw the vision as a reaction to population decline in Germany: Hamburg had to remain attractive to its citizens, draw young workers from other regions, increase tourism, and encourage businesses to relocate to the city in order to retain its international competitiveness. The vision is one that explicitly seeks growth, which is seen as the foundation for prosperity:

> The prerequisite for quantitative growth, which must precede qualitative growth, is to make the city so attractive that people actively choose Hamburg. Following the principle 'success breeds success,' Hamburg must be so nationally and internationally competitive and attractive that it regains global appeal. That is our goal, not out of vanity or a need for attention, but because we need growth to guarantee our children's future. (von Beust, 2004, 26)

Metropolis Hamburg includes four main goals: population growth, particularly of highly qualified migrants and families; economic and job growth through clusters with international appeal; increased land availability for housing and commercial development; and increased international attractiveness for Hamburg (FHH, 2002, 17ff). The document includes detailed strategies to achieve these goals.

In response to criticism from groups such as the sustainability-oriented *Zukunftsrat*, the vision's 2003 implementation document (*Fortschreibung*) enhanced the original plan's section on sustainability. Despite this, *Zukunftsrat* speaker Menzel complained that the plan's pursuit of growth bore no relation to sustainability principles (Menzel, 2004, 62). This vision and the implementation report were nevertheless received with open arms in political and administrative circles (Schindler, 2011).

The 2005 elections led to a new CDU–Green party coalition that modified the original growth vision to include sustainability in its new vision document, *Growth with Vision* (*Wachsen mit Weitsicht*), released near the end of its term in 2010. The new vision praised the success of the previous vision, but also used the recent financial crisis and acknowledgment of ecological limits to refine its understanding of growth:

> Conditions change constantly. This is evident in the current economic crisis: it is no longer enough to rely on the city's traditional strengths.... For this reason, it is important for Hamburg to open up new areas of development.... At the same time, unconditional growth is constrained by natural limits through increasingly scarce nonrenewable resources, and purely quantitative growth is often linked to not insubstantial external effects for the environment and society. Even growth that leaves some people behind does not meet our understanding of growth. For this reason, the Senate would like to engage more strongly in dialogue with these people. (BFHH, 2010, 1)

The Green-led vision maintained the goal of growth, but reframed it as linked not only to the economy, but also "equity and quality of life" (BFHH, 2010, 1). It included intentions to develop a sustainability strategy and put forward four main political goals for the city. Some, such as promotion of economic innovation and diversity and the goal of drawing talent to the city, remain from the previous vision, but green economy and social justice were added to these (BFHH, 2010, 2). As with the previous vision, *Wachsen mit Weitsicht* served primarily to explicate the government's political goals rather than as a blueprint for action, although it included five flagship projects. Two derived from the previous vision, and one of three new projects was called "sustainable Hamburg" (BFHH, 2010, 5). Sustainability was also called an overarching principle for action, and linked to the city's climate protection plan, its education for sustainable development initiative, and a balanced budget (BFHH, 2010, 5).

However, this vision was also rendered moot by political change: the 2011 elections brought in a SPD majority government. Indeed, current (SPD) mayor Olaf Scholz eschewed the idea of creating a new vision document after being elected,[2] saying that he cares more about concrete politics than visions (HH01; see also *SPD Regierungsprogramm*, 2011). The visions developed in Hamburg can thus be considered primarily political documents that have seen little resonance in the form of legally binding land-use and zoning plans. In Hamburg, the election

2 Scholz was re-elected at the beginning 2015, albeit without an absolute majority, and thus is likely to enter a coalition government with the Green party (Deutsche Welle, February 16, 2015).

cycle and partisan political landscape appear to play an important role in hindering the development and implementation of long-term strategic plans.

Despite this, Hamburg has been active in applying for outside funding and awards to dedicate to planning efforts and increase its reputation as a green city. In addition to winning a grant in 2001 in the German "regions of the future" (*Regionen der Zukunft*) competition, Hamburg was selected as the European Green Capital in 2011. This prestigious EU-level award seeks to promote environmentally sound urban living in part by funding environmentally oriented policy activity. In applying for the award, cities must demonstrate ambitious, ongoing environmental efforts and suitability as a role model for other European cities. Hamburg used its title to hold a series of workshops, which some hoped would form the basis of a sustainability report (HH01). Its status as Green Capital was also mentioned in the city's vision as a reason for Hamburg to serve as a role model for climate policy in Europe (FHH, 2010, 1). In this way, the city has made use of available outside funding to pursue sustainability planning activities: the workshops provide the basis for the city's environmental plan 2012–2015 (BSU, 2012). Here again, the city's environment ministry appears to be trying to take action aspects of sustainability under its purview. At the same time, it is unclear how effective BSU's efforts can be without being anchored in citywide activities.

This brief study of planning in Hamburg reveals a few key things. First, Hamburg has never had a citywide strategic plan guided by sustainability principles.[3] Instead, successive governing coalitions have developed vision documents to reflect their political goals, which have then been replaced by the incoming administration. Hamburg's vision documents have served more to clarify the current government's development approach and provoke political debate than to guide long-term planning.

At the same time, sustainability advocates have made numerous efforts to challenge unsustainable components of these visions or create a more targeted plan, as seen in the efforts of the environment ministry. Advocates have had some success in being heard and getting changes made at the margins (see Chapter 7 for a more detailed discussion). In addition, BSU's actions demonstrate some interest among political actors that could be better harnessed beyond environmental policy circles.

---

3 The literature represents conflicting accounts on this point. Although Schindler (2011) contends that Hamburg never had a comprehensive sustainability plan, Jörgensen (2012) argues that all German Bundesländer had passed a sustainability strategy by 2002, and Kern, Koll, and Schophaus (2007) call all three German city-states (Berlin, Bremen, and Hamburg) "pioneers" for implementing LA21s by 2006. More work remains to develop a more unified understanding of sustainability plans in the literature.

Table 5.2 Planning and sustainability in Hamburg

Plan	Key features	Link to sustainability principles
Metropole Hamburg – Wachsende Stadt (2002, 2003)	• Seeks to prevent urban population decline by making Hamburg attractive to middle-class workers • Includes economic clusters, education, transportation, and efforts to improve quality of life • Focus on international competitiveness	• *Acknowledges limits:* Uses economic and population growth as key goals • *Deliberative:* Vision developed by political elites without public engagement
Wachsen mit Weitsicht (2010)	• Acknowledges the previous vision and the need to adapt to changing circumstances • Acknowledges ecological limits	• *Acknowledges limits:* Tempers previous plan by acknowledging ecological limits • *Long-term:* Acknowledges need for long-term vision; plan abandoned after election of new mayor • *Systems-thinking:* Includes numerous agenda items, but limited links between them • *Reflexive:* Continuation of previous projects and readjustment of vision

Heidelberg: Consulting with the Public

In 1994, Heidelberg under mayor Beate Weber decided to develop a new strategic plan, as the most recent plan at the time had been created in 1974. The city determined that a new plan should include recent developments in the city and take environmental concerns into account (Stadt Heidelberg, 1997, 37). The project launched in October 1994 with a daylong event that included workshops linked to specific issues moderated by subject experts from institutions throughout Germany on issues such as "democracy and debate in the community" and "securing the future: the municipality's environmental responsibility" (Stadt Heidelberg, 1995).

The plan development process was led by the planning office and supported by city council committees, an advisory council composed of city council members, agency actors, and members of important local groups (such as architects, BUND, DGB, business groups, and the *Liga der freien Wohlfahrtspflege*) (ibid.). The advisory council discussed preliminary findings, particularly the administration's draft guidelines and goals (Stadt Heidelberg, 1997, 85). It commissioned empirica, an urban consulting firm, to provide support for plan development, particularly by conducting supporting background research (Stadt Heidelberg, 1997, 2002 [1997]).[4]

The firm empirica developed several reports as input for the decision-making process. The first was a SWOT analysis "that was particularly related to construction activity and the housing market" (Diefenbacher et al., 1997, 49; also, empirica and Stadt Heidelberg, 1995a). This analysis was released for public comment and sparked a wide range of criticism, particularly of its emphasis on housing and population/economic growth, which was seen as a hidden political agenda.

Critical responses from several political parties and organizations on the advisory council and the city council cautioned that the report seemed to take population growth as a given and see as the city's main goal provision of new homes within the city to families currently priced out of the market (for example, BUND, GAL in Stadt Heidelberg, 1997 [*Skizzen*]). The liberal democrats argued that a "scientific" study should have simply described development trends, the full range of possible goals, and potential instruments to achieve them, rather than provide its own goals. They were particularly incensed by the report's emphasis on creating new jobs: "Note: Right now there are 94,000 jobs in Heidelberg and 42,000 workers live here" (ibid.).[5] Several groups also criticize the report's implication that growth trends would continue (ibid.). Not everyone is critical in the same way, however; the response from the urban development agency agrees that high rents are a problem, but doubts that new construction is the solution, suggesting "further land-use policy measures and considerations for financing" (3).

empirica's second report discussed several development scenarios, essentially to answer the question "How big should Heidelberg become?" (HD05), but also to provide a blueprint for a "*zukunftsbeständige*"[6] urban development in Heidelberg: "a city that also offers a good quality of life to future generations" (empirica and Stadt Heidelberg, 1995b, 1). It presented three scenarios: status quo (in which no particular development objectives were presented), regional growth management (in which Heidelberg coordinated with neighboring municipalities to develop a regional development plan), and an inner development and local redistribution policy (in which land use was reduced through compact redevelopment) (empirica and Stadt Heidelberg, 1995b). After developing these scenarios, a fourth called

---

4 The citation used here attempts to clarify the unique publication history of this report. The original document was printed in 1997, but it was reprinted in 2002 (HD01). There is no date on the mayor's foreword or the second edition, which includes a new foreword by the mayor that refers to activities that happened after passage of the original plan, from which some of the information here is drawn. The dates therefore derive not from the document itself, but an interview conducted with a city administrator employed during the time in question.

5 Because the statements are provided in a bound volume with the SWOT analysis, there is no universal page numbering system. This quote comes from the first page of the liberal democrats' statement.

6 The term "*zukunftsbeständig*" indicates a weak form of "sustainability."

"City of Science Heidelberg" (*Wissenschaftsstadt Heidelberg*) was created to synthesize the regional growth management and redistribution options.

empirica then produced a short draft plan in November 1995, purposely kept short by omitting concrete policy proposals (empirica, 1995, [STEP], i). This was further developed by city government, experts, an advisory committee, and the city council and made available for public comment (Diefenbacher et al., 1997, 50; Stadt Heidelberg, 1997, 2002 [1997]). A second draft produced in November 1996 was used to guide public discussions on urban development in Heidelberg at five main events under the heading "Responsibility for the Future" and additional forums organized by various actors over a period of nine months. This discussion resulted in 360 proposed public amendments (by 10 individuals and 40 institutions) and 60 proposals by the city council (Stadt Heidelberg, 1997, 85; 2002 [1997], 2–3).

The various proposals were then discussed by the urban planning committee on 23 October 1996 and incorporated into an updated draft (Stadt Heidelberg, 1996 [Stellungnahmen]). Many were rejected; the changes consisted mainly of new clarifying blocks of text and fine-tuned language rather than major rewrites of the initial draft. Still, language was tweaked in the updated preamble to include mention of sustainable development and Agenda 21:

> The preamble puts as its first priority preservation of Heidelberg's individuality … and points to the goal of ~~economic success~~ *sustainable development*. Support of *Agenda 21 and* the Aalborg Charter brings … (Stadt Heidelberg, 1996, 8; strike-through and italics in original and indicate changes from previous draft).[7]

From the preamble of the 1997 plan:

> Heidelberg is striving to achieve a socially responsible, environmentally friendly and economically successful future development whilst at the same time it wants to retain its unique character. In doing so Heidelberg is guided by the goal of regional and global responsibility according to the Aalborg Charter. The key objectives of the 1974 urban development scheme serve as a basis (City of Heidelberg, 2002, iii).[8]

Heidelberg passed its urban development plan, after a brief delay caused by opponents in the city council, in 1997. The eventual vote of 31 to 2 was secured

7 This change was suggested by the Nicaragua-Forum (see Stadt Heidelberg, 1997 [Skizze], 15).

8 Heidelberg's development of this strategic plan gained the city international acclaim in the planning community, leading the city to translate several key documents into English in the early 2000s. Where possible, I have used the city's own English translations, marked by "City of Heidelberg" as opposed to "Stadt Heidelberg."

in part due to the overwhelming public resonance of the plan that resulted from the public discussions (Diefenbacher et al., 1997, 51), as well as the mayor's commitment to getting supramajority support (HD01). Soon after passage and in response to calls to develop a Local Agenda 21, the city commissioned environmental consulting group IFEU to analyze the city's actions with respect to sustainable development and Local Agenda 21 (Diefenbacher et al., 1997; IFEU and Stadt Heidelberg, 1997). The report concluded that STEP could serve as the city's Local Agenda 21 (IFEU and Stadt Heidelberg, 1997).

Heidelberg published its first report on implementation at the end of 2001, at which point it became clear that "the implementation of most of the targets pertaining to a sustainable city development would be ongoing tasks" (City of Heidelberg, 2004, 1). According to a 2002 summary of the 2001 report, "There is no separate chapter dealing with sustainability since this is an integral part of the City Development Plan as well as all its chapters" (City of Heidelberg, 2002, 4).

The process of participation used in Heidelberg was an effective hybrid of direct public consultation and expert-driven planning. Although the public comment periods gave major stakeholders an opportunity to be heard in the planning process, the original draft by empirica, a firm harshly criticized in early stages, was adopted with generally cosmetic changes. This demonstrates both the power held by those brought in early in the process, as well as the extent to which engaging the public can increase plan support, even if many of the public's ideas are not incorporated.

In 2006, under mayor Eckart Würzner, the city updated its development plan to 2015 with the unanimous support of the city council (Stadt Heidelberg, 2006). Although the updated plan added a new chapter on demographic change, it did not change the 11 existing chapters (Stadt Heidelberg, 2006). This updated plan shows longevity of the vision behind the 2010 STEP, but also suggests planning fatigue among decision-makers. Despite being realized nearly 10 years after the previous strategic plan, it only extends 5 years into the future. In addition, the lack of change to the updated plan's content may indicate a lack of reflection on new developments and changes. There is currently a discussion in the planning office on how to update the STEP, but the new plan will not be ready until at least 2014 (HD01).

After I conducted my field research, the city released a document projecting the city's housing needs in 2030 (Stadt Heidelberg and GEWOS Institut, 2013). This document reflects the growth mentality of city decision-makers seen also in Berlin, Hamburg, and New York that was challenged by the public in all of these cities, as well as during development of Heidelberg's own STEP process. It will be interesting to see how this report affects future planning in Heidelberg, particularly as the city's strategic plan is updated.

Planning in Heidelberg has included some level of public participation and institutionalized sustainability principles—as measured by STEP goals—into urban decision-making. At the same time, it appears that since Mayor Weber left office, sustainability has lost the momentum it enjoyed under her administration. As one of my interview partners told me, "Sustainability in Heidelberg is administered, not lived" (HD04). Although the overwhelming majority that passed the city's STEP has led to its continued use, the plan will likely not be updated before 2015. This makes it unlikely that Heidelberg's plan is still fulfilling its role as a meaningful planning instrument. In addition, institutionalized mechanisms such as the sustainability test may be less effective than expected in an administrative climate that lacks real enthusiasm for sustainability issues (an issue explored further in Chapter 6). Here, the city's motivation to actively engage with sustainability appeared to be tied to a dynamic leader. Thus, although sustainability principles have nominally remained part of city decisions since Weber's original planning efforts, the concept seems to have lost its luster in the ensuing years, signaling the limits to sustainability planning that relies on an institutional component alone.

Table 5.3 Planning and sustainability in Heidelberg

Plan	Key features	Link to sustainability principles
Stadtentwicklungsplan Heidelberg 2010 (1997)	• Extensive public engagement process increased plan acceptance, but some worries that initial consultants' reports over shaped plan toward growth outlook • Acknowledgment of limits, UN debate	• *Acknowledges limits:* Yes, with language linking plan to global sustainability concerns • *Long-term:* 15-year planning horizon • Systems-thinking: • *Deliberative:* Strong stakeholder and public engagement mechanisms
Stadtentwicklungsplan Heidelberg 2015 (2006)	• Includes new chapter on demographic change	• *Long-term:* Nine-year planning horizon the shortest of the plans studied here; 2015 plan still in place in summer 2013 • *Reflexivity:* Little change in update from 2010 plan

New York: A Stakeholder-based Approach to Sustainability

Sustainability has a relatively short history as a political concept in New York, and exists primarily through the initiatives promoted in the city's sustainability plan, *PlaNYC 2030*, released in 2007. In this case, the city's strategic plan became a way to introduce sustainability concepts and values into its decision-making. However, *PlaNYC* did not originate as a sustainability plan, and its origins reflect its purpose in city decision-making.

In early 2002, recently elected Mayor Bloomberg hired Dan Doctoroff as deputy mayor for economic development. Doctoroff had founded NYC

2012, which had developed a proposal for New York City's 2012 Olympic bid that envisioned redeveloping several large, unused tracts of city land and revitalizing numerous neighborhoods throughout the city (Bagley and Gallucci, 2013). Doctoroff was asked to think more strategically about the city's built environment, something that had not been done for decades in a city facing constant crisis.[9] Bloomberg asked Doctoroff to create a long-term economic and infrastructure development plan to realize his Olympic vision even if the city lost its bid (Bagley and Gallucci, 2013).

The plan was to rejuvenate the city's infrastructure and create space for nearly 1 million new residents expected by city projections. As the plan took shape, however, city officials became aware of the scale of potential future problems caused by a growing population on the city's aging infrastructure (many of its water and subway tunnels were over 100 years old). New housing developments would require additional transportation infrastructure and increase water use. Much of the city's infrastructure had been cutting-edge in the 19th century, but had been underfunded for decades and would be unable to accommodate 900,000 new inhabitants. Therefore, the city had to dramatically improve its aging infrastructure if it wanted to capture anticipated growth without sacrificing quality of life.

During this time, the city was also developing a greenhouse gas inventory and drafting policies to reduce its emissions. It had established partnerships with researchers at NASA's Goddard Institute for Space Studies who were conducting extensive analysis on issues like sea level rise, precipitation, and forecasts on the number of hot days (Bagley and Gallucci, 2013; NY11). This work on climate change highlighted the city's main emissions culprit: buildings. Retro-fitting old buildings and designing new ones to be more efficient required work that dovetailed nicely into the economic development team's infrastructure planning. Further work on the plan highlighted how complex and cross-cutting infrastructure issues were, as well as their connection to the city's work on climate change. Joining forces would allow the city to realize both economic and climate goals simultaneously.

So in January 2006, Mayor Bloomberg asked Doctoroff "to begin a citywide conversation about sustainability" (Rosan, 2012). In order to coordinate this

9 New York is a classic example of a Northeastern industrial city, most of which saw population and economic decline with the growth of automobile use and racially driven suburbanization. In the 1970s, the city faced bankruptcy. Tax revenues were at new lows as the wealthy and middle-class left the city, making it impossible to provide social services to the urban poor who remained. In the succeeding decades, New York City mayoral policy focused on averting further crisis by trying to balance the city budget, reduce crime, and make the city attractive to suburbanites and business. This pattern is emblematic of many Northeastern U.S. cities and well documented by Beauregard (2006), Rybczynski (2007), and others.

effort, the Office of Long-Term Planning and Sustainability (OLTPS)[10] was created in September 2006. It was tasked with coordinating plan development in conjunction with consultants from McKinsey, staff from various city agencies, and an external Sustainability Advisory Board composed of 16 major stakeholders from the nonprofit and private sectors (City of New York, 2007). The advisory board included professors, representatives from national environmental NGOs and local environmental justice groups, and business leaders (Naparstek, 2006).

Over the next 18 months, the advisory board held countless closed-door meetings as well as several public events to further develop the plan. While board members acknowledged that public input was "truncated" due to the ambitious schedule the term-delimited mayor[11] had for implementation (NY08), opponents argued that the plan was "monologue, not dialogue" (Angotti, 2008). The process described in Bagley and Gallucci (2013) is one in which key elements of the plan were developed at the board's off-the-record meetings with city administrators (Loc. 450ff.); evaluation of participation also differs markedly between those included in the process through membership on the advisory board, and those excluded from official channels (for example, Craytor in Bagley and Gallucci, 2013; NY08; NY10). The process can thus be described one of intense engagement of a limited number of key stakeholders, with far less involvement beyond that group. It is clear that the mayor's office felt that a more inclusive process would have taken extra time that it did not want to spare (NY08; NY09). Indeed, then–OLTPS director Rohit Aggarwala doubted that the sort of grassroots inclusion some advocates longed for was even possible in a city of over 8 million inhabitants (DePillis, 2008).

PlaNYC 2030 was released in April 2007 (City of New York, 2007; ICLEI and City of New York, 2010). The plan, which focuses on the physical city, included 127 initiatives related to 10 issue areas, including housing, parks and public space, waterways, energy, air quality, climate change, and congestion, with the goal of making the city a leader in sustainability by 2030. The plan's release, a little over a year before the real estate and financial crisis hit the United States, led to ebullient forecasts in revenue and population growth. The recession meant a mild decline in revenue from forecasts, thus reducing available funding for many initiatives. However, reductions in the capital budget did not seem to negatively affect most initiatives; the 2009 progress report listed delay of 28 percent of milestones and "redirection" of a further 5 percent in light of new information (City of New York, 2009).

PlaNYC 2030 received substantial press and publicity soon after its release,[12] including a major event on the plan hosted by the Institute for Urban Design

10 Bagley and Gallucci (2013) present the office as opening during July 2006 without the "sustainability," which came later when director Rohit Aggarwala saw sustainability as the connecting thread between all the plan's initiatives (Loc. 350).

11 Bloomberg later got the city council to pass legislation allowing him to run for a third term, which he won, allowing him to serve through December 2013.

12 Also internationally, such as in the *Frankfurter Allgemeine Zeitung* (Schwengler, 2014).

that included key stakeholders as well as major critics (Byles and Kazi, 2008). The mayor's interest in PlaNYC, and his creation of OLTPS within the mayor's office, has led to important partnerships with nonprofits, businesses, and outside agencies, as well as improved communication and coordination between city agencies and thus strengthened plan implementation. The initiatives that have leveraged public–private partnerships, such as the Million Trees initiative and a new bike rental scheme co-sponsored by Citibank (Flegenheimer, 2014),[13] could inspire other cities facing resource constraints. Such coordination has allowed the city to move forward with a massive revitalization process that would have been unlikely to succeed without the strategic framework provided by PlaNYC.

Although the 2011 plan update added waste management goals, PlaNYC is primarily an infrastructure plan focusing on the city's built environment. As such, it is not a truly strategic plan, but still a major planning initiative for a city with no planning culture since Robert Moses in the mid–20th century. The plan is results-oriented, including specific initiatives with policies and short- and medium-term targets. It includes innovative ideas, such a plan for congestion pricing for cars entering lower Manhattan during peak times that would be used to fund public transportation, although this particular initiative failed due to state-level opposition (Bagley and Gallucci, 2013).

Since the plan's initial publication, OLTPS has released annual progress reports on plan implementation as well as a plan update in 2011 that adds waste management issues, which were excluded from the original plan because of parallel work by the city's Department of Sanitation on a Solid Waste Management Plan (NY08).[14]

The largest hurdle to the plan's goals came for initiatives over which the mayor lacked direct control: congestion pricing and hybrid taxis. Perhaps the plan's most visible initiative was its congestion pricing scheme, which would have fully funded the city's transit infrastructure by introducing a fee to enter Manhattan's main commercial district on weekdays (City of New York, 2007). This initiative was rejected after the New York state Senate refused to bring it to a vote in 2008 (Bagley and Gallucci, 2013), exacerbating the already tense relationship between the city and the state. In addition, the plan to require all taxicabs to become hybrid was challenged by the NYC Limousine and Taxicab

13 Citibank's sponsorship of Citi Bike, the city's bike sharing program, has led to problems, though. The bank's logo through the bike program's branding has made it difficult to find other sponsors; several docking stations have been vandalized by critics of Citibank (Ferro, 2014).

14 This was due to parallel work on a new Solid Waste Management Plan, a notable attempt to deal with New York's historically low recycling rates. The city's waste management is complicated in part to the strength of waste collection unions with Mafia ties (NY01). The SWMP is also remarkable because it calls for the first time for waste transfer stations to be sited in each of the five boroughs, equalizing their negative impacts, a triumph for environmental justice advocates, who form the core of the city's environmental movement (Rosan, 2012). However, lawsuits raised by residents in wealthier boroughs, such as Manhattan, throw into doubt whether the original plan will take effect as intended (NY01).

Commission and lost in the Supreme Court, which found the city lacked jurisdiction for such legislation (Gross, 2011).[15] The failure of these initiatives highlights the importance of state and federal government action, which can be seen as an important constraint on New York's initiatives, as well as getting the support of key actors early on. Where stakeholders were brought into the process, such as through the Sustainability Advisory Board, they became supporters of the plan and its initiatives. It also raises the question of what role the press plays in framing sustainability issues, as more successful initiatives have seen far less coverage than the political struggles.

Indeed, lack of media attention may also be related to a lack of public visibility of the plan, which has failed to spark broad interest or engagement beyond the early controversies listed above. PlaNYC has been less prominent than Mayor Bloomberg's more controversial policies, such as smoking and soda bans, gun control, and climate change (Bagley and Gallucci, 2013). This may be related to the New York press's confrontational stance: because the plan was generally well regarded, there has been little reason to publish further on successful—and conflict-free—implementation.[16] Perhaps a larger problem is that the plan has not been institutionalized into city decision-making processes with a legal mandate for implementation,[17] leading to fears about what will happen when Bloomberg leaves office in 2014 (NY03; NY05). The lack of public advocacy for the plan or sustainability, as well as PlaNYC's close identification with Mayor Bloomberg, make it unclear the extent to which the plan's initiatives will continue under the new DeBlasio administration.

The 2011 plan update replaced a section on "state of good repair" for all subways, roads, and rails with one dealing with solid waste, responding to the failure of congestion pricing as well as the importance of better waste management to the city and its sustainability goals. In addition to the performance measures linked to plan initiatives (see Chapter 7), the plan's approach has been very data-intensive, encouraging partnerships with environmental groups and local universities that collect new information that aids decision-making. For example, a lack of solid

---

15 Subsequently, the city held a competition for a "Taxi of Tomorrow" to unify the city's massive taxi fleet, eventually choosing a winner manufactured outside the United States whose fuel efficiency was lower than the existing fleet average (Gross, 2011).

16 In addition, a major theme that emerges from Bagley and Gallucci's (2013) book on PlaNYC is the importance of media secrecy to plan developers. Throughout, members of the administration and the sustainability advisory board emphasize the problems that would arise if the media were to hear about the issues they were discussing, such as congestion pricing. Such secrecy necessarily narrows the scope of the deliberative process; one wonders what would have happened with such initiatives if board members had been more open.

17 Aspects of the plan have been incorporated into city law, however, such as energy efficiency and pollution requirements for buildings, the city's climate change panel, which "is legally required to update its global warming projections at least every three years," as well as the city's work on climate resilience (Bagley and Gallucci, 2013, Loc. 1438f.).

data on street-level air quality led the city to create an initiative to "launch [a] collaborative air quality study" to help meet its goal of having the cleanest air in a major U.S. city (City of New York, 2007). Although originally created with an emphasis on car and truck emissions, the 150 street-level monitoring stations revealed an unexpected culprit: the heating oil used in 10,000 older Manhattan buildings cause the same amount of air pollution as all of the city's vehicles (Bagley and Gallucci, 2013; NY02). These data led to creation of legislation that phases out their use by 2015 and 2020, respectively. This example shows how the city began with a goal—clean air—and then collected data to meet that goal, revising policy as necessary.

Sustainability in New York is clearly connected to a long-term vision of the city that connects various policy sectors in order to coordinate policymaking in a more strategic way. It focuses on bringing the city's infrastructure up to 21st-century standards and uses innovative mechanisms, such as public–private partnerships, to do so. Although it is primarily an infrastructure plan that emerged from economic development goals, PlaNYC links initiatives to multiple issue areas. For example, one initiative under its water rubric calls for a water conservation program to reduce citywide consumption by 60 million gallons per day, and is linked to water quality, water network, energy, and climate change goals (City of New York, 2007, 67).

However, although the city links its understanding of sustainability to a sense of global responsibility on issues like climate change, it is also clearly linked to a growth paradigm. PlaNYC frames its initiatives as necessary to maintain global competitiveness; housing projects tend to focus on attracting internationally talented workers rather than dealing with the housing issues faced by current residents. Although the plan arose from projections that saw the city's population rise by 1 million residents, the plan appears to encourage and spur such growth rather than accommodate it. This may be due at least in part to its history of wealth creation from real estate development, regional and international trade, and financial services, but it demonstrates that sustainability may be used to some extent to soften the hard edges of the city's growth agenda.

Much of the underlying motivation for action has been economic. In addition to starting as a comprehensive land development plan, many of PlaNYC's initiatives have an implicit or explicit goal of creating economic growth or saving the city money. The director of the MillionTrees project told me that an EPA-supported analysis of trees that valued a $1 investment at over $5 was "instrumental" in getting the mayor to support the initiative (NY03). While some may argue that this could weaken the city's definition of "sustainability," it is just as likely that economic language could motivate actions that may not otherwise be taken. For this reason, it may be worthwhile to explore such accounting, at least in the short-term, to sell sustainability initiatives.

Table 5.4 Planning and sustainability in New York

Plan	Key features	Link to sustainability principles
PlaNYC 2030 (2007, updated 2011)	• Originated as a development plan to accommodate projected population increase • Includes concrete initiatives linked to built environment and climate change components • High-level mayoral support has aided implementation during Bloomberg administration, but lack of institutionalization	• *Acknowledges limits:* Refers to climate change, but to no other ecological limits • *Long-term:* Especially long time horizon for New York decision-making • *Systems-thinking:* Connections between sectors, and development and climate change • *Deliberative:* Strong stakeholder deliberation with Sustainability Advisory Board, but limited engagement with the public and the media • *Reflexivity:* Policy modifications in 2011 plan update; potential inclusion of equity in 2015

Portland: A Comprehensive Plan by the People

A city with a strong culture of planning, Portland has been subject to state-mandated planning requirements since the 1970s (Wheeler, 2003).[18] The state of Oregon, in which Portland is the largest city, requires that cities and counties create Urban Growth Boundaries (UGB) to restrict land development to targeted areas. The policy has two purposes: to protect high-quality farmland near the city from development and to support a compact urban form. These purposes were reflected in the early coalition of farmers and city-dwellers who banded together to pass the original statute and have supported it in the decades since (PDX01; PDX05). Portland's UGB, created in 1979, has contributed to the revitalization of its downtown, as well as mitigated the sort of car-dependent sprawl that Oregonians dread (Richardson and Gordon, 2001; PDX03).

In 2012, Portland was in the final phase of a three-step comprehensive planning process. This process began with VisionPDX, a three-year community visioning process, followed by development of the *Portland Plan*, a strategic plan passed unanimously by the city council in 2012, and implemented through updates to the state-mandated comprehensive plan. Oregon state law requires a "periodic review" of city comprehensive plans to ensure that they are consistent with statewide planning goals (DLCD, 2007). The statute includes participation requirements

18 SB 100, the original statute requiring comprehensive land-use plans, was advocated for by then-mayor Goldschmidt (Abbott, 2004; PDX01), who was an important advocate for the Urban Growth Boundary (PDX05).

and development of a work program detailing the city's review process. Oregon's Department of Land Conservation and Development (DLCD) also provides grants to complete this work program (DLCD, 2007). In this way, the state provides not only a mandate for planning, but also requirements on process and content and resources for development and implementation.

The first step of Portland's recent plan update was to develop a vision for the city. VisionPDX was a massive multi-year visioning project conducted from 2005 to 2007 under Mayor Tom Potter. The initiative was co-chaired by a dean at Portland State University, as well as the director of the university's metropolitan studies institute. The process sought broad public input on Portland's future in order to develop Portland residents' vision for the city.[19]

The visioning process was guided by a 57-member volunteer Vision Committee that represented a broad cross-section of stakeholder groups. The process, according to the report, aimed to create ownership within the community rather than simply "buy-in" (VisionPDX, 2008, 42). Thus, efforts were made to establish relationships within the community before asking for input on the survey or in interviews; grants to community groups helped with this and collecting survey responses. The report cites "long-lasting impact" as groups learned about the communities they served, formed partnerships, and taught participants and volunteers leadership and organizing skills (VisionPDX, 2008, 44).

The committee used an open-ended survey with four questions: 1) what do residents value most about their community, 2) what changes they would like to see, 3) their vision for Portland in 20 years, and 4) how to get there (VisionPDX, 2008, 6, 43). Over 21,000 pages of comments were collected. The results were published as *Portland 2030: A Vision for the Future* (VisionPDX, 2008). To emphasize the level of public consultation, the report claims to be authored "by the people of Portland, Oregon" (VisionPDX, 2008). It frames the plan's work to address challenges of growth, increased diversity, climate change and peak oil, and economic change.

The results of VisionPDX formed the foundation of the *Portland Plan*, the city's strategic plan that was developed from 2009 to 2012. The planning process had four phases: conducting background research to determine the city's current state; drafting plan goals and objectives; developing draft integrated strategies; and presenting the three strategies for public review (City of Portland, 2012, 142–3). Although

19 Mayor Potter, in his introductory letter to the vision document, described his understanding of participatory engagement: "When I commissioned this project, I wanted to ensure that all Portlanders had an opportunity to participate in this important discussion – particularly those who had not previously been asked to sit at the public policy table. Since then, I have watched as City Hall filled with hundreds of immigrants and refugees from all over the world who took their voices into Council Chambers to be heard. Our small business community produced a report with hundreds of ideas to encourage local economic development. Students involved in one of our visionPDX grants helped organize and pass a Bill of Rights for young people. Now there are youth conducting research on how they can be better engaged and served in our city" (VisionPDX, 2008, 1).

The Portland Plan retains many of the same themes and priorities, it differs from VisionPDX in a few ways. First, the plan advances the target year from VisionPDX's 2030 to 2035. It also uses what it calls "a framework for equity" to focus on three goals: economic prosperity, quality education, and human and environmental health (City of Portland, 2012). Instead of focusing on the built environment like most other plans studied here, *The Portland Plan* focuses on meeting people's needs through equitable, cost-effective, and well-coordinated service provision. *The Portland Plan* places less emphasis on the term "sustainability," but its long-term initiatives on education and complete neighborhoods[20] are consistent with sustainability principles. Environmental objectives take a back seat to the more pressing concerns of Portland's residents, but as the plan says, it is "a plan for people" (City of Portland, 2012, 3). Indeed, the reduced emphasis on traditional areas of urban planning may be due to the city's strong planning history and well as the larger role in planning played by Metro. This stands in contrast to a city like New York, where city officials sought to tackle public safety and educational issues before focusing on the built environment with *PlaNYC*.

The Portland Plan was passed unanimously by city council in April 2012. Because it was adopted by resolution, it is not legally binding and thus only "serve[s] as a guide for future government action" (BPS, 2013b). Implementation thus takes the form of incorporation into the state-mandated comprehensive plan, which is being completely updated for the first time since it was passed in 1980, as required by the state's periodic review (which "requires that a comprehensive plan have sufficient and effective implementing measures"). As of July 2013, the first section of a draft comprehensive plan that "includes initial draft goals and policies for public discussion and review" has been released; a second part "will include draft maps and a draft list of capital projects" (BPS, n.d.). The comprehensive plan has already added one goal to its subtitle in addition to those from the *Portland Plan*: resilient (BPS, 2013a). This provides more evidence of the subtle changes that occur even in a highly orderly planning process like Portland's, as well as perhaps reflexivity in implementation.

It is clear that Portland's process, aided by state comprehensive planning requirements, led to a far-reaching strategic plan with broad public support and a better chance of implementation. But to what extent does Portland's plan incorporate sustainability principles into decision-making? Sustainability was explicitly listed as one of the three core values distilled in VisionPDX's 2008 visioning document and is used in numerous ways in both the strategic and

20 In terms of built environment, one of the plan's goals is for 80 percent of residents to live in "complete neighborhoods," which meet needs of all residents for housing, food and commercial services, education, open space and recreation, active transportation options ("it is built at a walkable and bikeable human scale"), and civic amenities (City of Portland, 2012, 76).

comprehensive plan.[21] However, it was not an explicit goal during the plan process as in Seattle (see below).[22] At the same time, some writers have argued that true sustainability planning is that which involves the local community in a collaborative process (Innes and Booher, 2010; Wheeler, 2004). The length of the entire process demonstrates the extended time and resource commitment for such extensive public engagement, however, especially when contrasted with a plan like New York's, which was developed and implemented in less than two years. Equity, participation, and social cohesion emerge as guiding principles for the cities, which was a departure from the values of growth and international competitiveness that drove other plans.

Although regional planning is beyond the scope of this book, it is worthy of brief mention in the case of Portland. In addition to the city's Bureau of Planning and Sustainability, Portland is home to Metro, the only elected regional government in the United States. Metro's main task is regional development, with the city's Urban Growth Boundary as a key policy instrument under its purview (PDX07; Seltzer, 2004). Metro created a Region 2040 Growth Concept to guide development in 1997 (Wheeler, 2003), and uses 50-year urban and rural land reserves to aid long-term planning (PDX07). Metro works closely with the city's planning agency and other administrative bodies: employees at the regional and city level keep in regular phone contact (PDX04). This demonstrates close collaboration on planning issues and the city's long-term view.

Portland's recent comprehensive plan update was perhaps the most inclusive and certainly the longest process of the six cities studied here. Although the earlier vision's reference to sustainability has been replaced by an equity framework, the plan's language and the city's planning culture reflect many of its principles. Portland's experience shows the important role state planning requirements can play in plan development and inclusion of sustainability principles. In addition, it

21 For example, the Portland Plan seeks to "enhance Portland as a national model for sustainability and as a center for business development by commercializing sustainability practices, products and services," and also highlights the city's active transport, green stormwater, and open space systems (City of Portland, 2012, 54). It also explicitly links sustainability to efficient freight infrastructure and efficiency processes in businesses, including green procurement strategies. The comprehensive plan links sustainability to urban innovation and the city's business strategy, as well as "environmental sustainability" in the sense of reduced resource use and efficiency gains. Sustainability is listed as a goal for transportation projects: "prioritize projects that will reduce vehicle miles traveled per capita and increase the green and active transportation mode shares while increasing access to opportunity and supporting health, economic vitality, sustainability, wise application of resources, and community values" (BPS, 2013a, 7–19, my emphasis).

22 It should be noted, however, that the office responsible for comprehensive planning in Portland is the Bureau of Planning and Sustainability. This office is the result of a merger between the city's Office of Sustainable Development (OSD) and its planning office under Mayor Adams to better link sustainability goals to planning (PDX06).

suggests the role that a deliberative process may have in moving the plan's goals from growth to quality-of-life.

Table 5.5 Planning and sustainability in Portland

Plan	Key features	Link to sustainability principles
VisionPDX (2008)	• Inclusive public input process to develop city vision • Climate change and peak oil seen as challenges	• *Acknowledges limits:* Efforts to preserve scarce farmland • Considers complexity by using multiple connected thematic areas • Strong deliberative component
The Portland Plan (2012)	• Uses equity framework • Focuses on services rather than built environment • Public responses for *VisionPDX* guide plan structure and content	• *Acknowledges limits:* Efforts to preserve scarce farmland • *Long-term:* Connected to long-term regional planning (40 years) • *Systems-thinking:* Considers complexity by using multiple connected thematic areas • *Deliberative:* Strong deliberative component • *Reflexive:* Linked to planning requirements

Seattle: Strategic Sustainability Planning

Sustainability planning in Seattle began in the early 1990s, spurred by state legislation that required cities to develop comprehensive growth management plans. The city has used a broadly participatory approach but, unlike Portland, explicitly uses sustainability as a framework for its state-mandated comprehensive plan, *Toward a Sustainable Seattle*. Despite this, it appears that Seattle's plan serves more as an almanac of policies passed elsewhere than a guiding document for city policymaking.

Seattle was an early adopter of sustainability in its urban planning, releasing its first comprehensive plan shortly after the first Rio summit on sustainable development, in 1994. This plan was mandated by the *Growth Management Act* passed by Washington state in 1990.[23] The new law, modeled on Oregon's successful growth management legislation (Lloyd, 2000; Miller, 2004), required all major or fast-growing cities

23 The Act's original preamble explicitly aims to preserve "the environment, sustainable economic development, and the health, safety, and high quality of life enjoyed by residents of this state," which it sees as threatened by "uncoordinated and unplanned growth" (Lloyd, 2000, 83–4).

and counties to develop comprehensive plans that concentrated growth into transit-accessible urban areas and that protected sensitive environmental and rural resource land (Laschever, 1998; Lloyd, 2000). The act was based in part on King County's 1985 comprehensive plan (Miller, 2004) as well as the recommendations of Seattle's Economic Development Commission (1990). This was nothing new; Seattle had growth management policies as early as 1977 (Brambilla and Longo, 1980).

As suggested by state law, Seattle has focused in its planning on growth management: accommodating new residents while preserving the city's unique character and high quality of life. Seattle mayor Norm Rice devoted significant resources to developing the city's comprehensive plan, holding several town hall meetings and running the plan's development as a political campaign, which meant both soliciting broad public engagement as well as carefully framing the plan in a way that garnered support (SEA05). However, this process was used as much to accommodate opponents as to pursue political strategy: early opposition to neighborhood growth targets upon release of a draft plan forced city planners to cede power back to the city's strong neighborhoods. To accommodate the neighborhoods, the city plan uses an urban villages model that provides population and jobs targets for each neighborhood, which is then asked to create a neighborhood plan using a resident-driven process that includes a neighborhood planner and resource support from the city's Department of Neighborhoods, allowing them to determine their own strategy for growth based on their neighborhood's particular character (CSC et al., 1995). This approach improved political support for the comprehensive plan, but has been criticized as allowing powerful private citizens—primarily middle- and upper-class homeowners—to drive the planning agenda (SEA01). In addition, according to one planner involved in the process, including the public is difficult, especially on abstract issues such as sustainability, where planners "get a lot of blank stares" from non-experts (SEA01). Still, linking neighborhood plans to the citywide plan has helped neighborhoods select local infrastructure projects such as libraries (Office of City Auditor, 2007).

The plan was updated in 2004 to deal with development to 2024, and a background report was released in early 2014 to begin an update to 2035 (DPD, 2014). It remains based on four core values—community, environmental stewardship, economic opportunity and security, and social equity—that it links to four dimensions of sustainability (DPD, 2005; DPD, 2014). The plan includes numerous "elements," many of them mandated by state law, but also adding the new elements "Human Development" and "Environment" after passage of the initial plan (DPD, 2005). Still, the plan's main purpose is growth management:

> The current [i.e., 2005] Plan calls for the City to embrace growth. Today our challenge is a different one—how can we leverage growth to build better neighborhoods, create jobs and economic opportunity for all our residents, and improve the safety and vitality of our city? (DPD, 2014)

The four core values thus serve to describe how the city is to approach the growth challenge, informing a broader vision of how growth should be accommodated in a way that reflects those values. Although it includes additional elements beyond the state requirements, the plan is essentially a strategy for land-use development, focusing on how to accommodate projected population and job growth. These targets help city administrators plan infrastructure developments such as parks and transportation links (SEA02). At the same time, the plan serves as a sort of almanac of policies related to planning in the city, as state-required amendments on issues such as transit, urban design, climate action, and healthy food (the 2012 themes) are added to the plan each year (SEA02).

A relatively orderly political process as well as the statutory requirement for a city comprehensive plan have led to the plan's longevity in Seattle. However, the extent to which the plan serves as a roadmap for development is unclear. While planners see the value of *Toward a Sustainable Seattle*, others have argued that the plan's major function is as a compendium of urban policies passed through other channels (SEA03). Thus, Seattle's plan should be viewed more as an effect of the city's interest in sustainability than a source of new policy action. Still, it did give the city an opportunity to start a discussion about integrating sustainability principles into development of the city's built environment.

One academic claimed that Seattle is the only city in the United States that takes a holistic, social view of sustainability (SEA04). The city's approach came from a mix of the grassroots engagement and state-level planning requirements. In addition, several community groups have been engaged, such communities of color, the poor, and social justice advocates, because they see a common agenda. Indeed, grassroots and civil society engagement often aided the planning process (see Chapter 7). Planning in Seattle enjoys the efforts of motivated actor groups inside and outside government who have ensured the plan's continued salience. At the same time, more work with neighborhood groups could improve implementation at that level.

Table 5.6 Planning and sustainability in Seattle

Plan	Key features	Links to sustainability principles
Toward a sustainable Seattle (1994, 2005)	• Based on core values of community, environmental stewardship, economic opportunity, and social equity • Focused on accommodating population growth in a way that reflects values and maintains quality of life • Primarily a built environment plan; updated by state mandate every 10 years • Inclusion of neighborhood plans devolves some planning and increases engagement	• *Acknowledges limits:* Growth management frame acknowledges limits of social services and infrastructure • *Long-term:* 20-year timeframe • *Systems thinking:* Links transportation, land-use, and jobs • *Deliberative:* Attempts at citizen engagement with focus on neighborhood empowerment • *Reflexive:* Has been updated regularly; updates part of state mandate

Analysis: Strategic Plans and Urban Sustainability

Efforts were seen in all six cities to use planning instruments to promote sustainability. Several, such as Hamburg and Berlin, even tried different types of plans, such as vision documents, urban development plans, and Local Agenda 21s. Plans can thus serve a variety of functions, such as sparking public debate about the city's future, presenting a political vision for the city, linking initiatives across policy sectors, and developing partnerships between local stakeholders. A plan is not a specific policy, but a framework to guide policymaking across sectors by providing strategic guidelines and policy priorities.

Plans are by nature forward-looking. They provide a unique opportunity to integrate sustainability thinking into a city's long-term vision and development. However, this does not mean that creation of a plan makes a city sustainable. Some of the plans studied here made little or no mention of sustainability; others used it instrumentally to achieve other goals, often related to growth. Simply developing a strategic plan or vision does not mean that it adheres to sustainability principles, even when the term "sustainability" is used.

This analysis focuses on two issues: why cities develop strategic plans and how these plans can incorporate sustainability principles into political decision-making. Although each case studied here represents a unique experience defined by numerous contextual factors, some important lessons and challenges emerge.

Table 5.7 Strategic plans and sustainability in six U.S. and German cities

City	Key features of plan and process	Incorporation of sustainability principles
Berlin	• LA21 developed by engaged external actors; no follow-through • Supracity funding promoted LA21 process, but meant initiatives dried up without it • New urban development plan created through support of planning minister; no linkage to city's LA21	• *Acknowledges limits:* LA21 uses sustainability framework; Berlin2030 through discussion of climate neutrality • *Long-term:* 25-year outlook in first plan; 15-plan in second • *Systems-thinking:* LA21's cross-cutting themes acknowledge complexity; Berlin2030 acknowledges conflicts within and between goals • *Deliberative:* With LA21, achieved through thematic working groups and engagement of civil society actors; Berlin 2030 had limited public engagement, but calls for participatory planning • *Reflexive:* No acknowledgment of LA21 in new plan
Hamburg	• Visions developed by political coalitions • Growth orientation challenged by *Zukunftsrat*	• *Acknowledges limits:* Original vision uses economic and population growth as key goals; revision tempers previous plan by acknowledging ecological limits • *Long-term:* Acknowledges need for long-term vision; plan abandoned after election of new mayor • *Systems-thinking:* Includes numerous agenda items, but without explicit links between them • *Deliberative:* Vision developed by political elites without public engagement or inclusion of opposition parties • *Reflexive:* Continuation of previous projects and readjustment of vision in revision
Heidelberg	• Plan developed through massive deliberative process • Currently merely "administrated": needs update	• *Acknowledges limits:* Yes, with language linking plan to global sustainability concerns • *Long-term:* Initial 15-year planning horizon; nine-year planning horizon after update; 2015 plan still in place in summer 2013 • Systems-thinking: • *Deliberative:* Strong stakeholder and public engagement mechanisms • *Reflexivity:* Little change in update from 2010 plan

New York	• Plan originated as economic development • Developed by elite major stakeholders	• *Acknowledges limits:* Refers to climate change, but to no other ecological limits • *Long-term:* Especially long time horizon for New York decision-making • *Systems-thinking:* Connections between sectors, and development and climate change • *Deliberative:* Strong stakeholder deliberation with Sustainability Advisory Board, but limited engagement with the public and the media • *Reflexivity:* Policy modifications in 2011 plan update; potential inclusion of equity in 2015
Portland	• Developed with intense public input process • Uses equity framework • Aided by state regulation, support	• *Acknowledges limits:* Attempts to preserve scarce farmland, restrict growth to within UGB • *Long-term:* Connected to 40-year regional planning • *Systems-thinking:* Considers complexity by using multiple connected thematic areas • *Deliberative:* Strong deliberative component • *Reflexive:* Linked to planning requirements; refined deliberative approach to reach underrepresented groups
Seattle	• Aided by state regulation • Strong initial neighborhood involvement, but lack of institutional follow-through • Four core values closely related to sustainability	• *Acknowledges limits:* Growth management frame acknowledges limits of social services and infrastructure • *Long-term:* 20-year timeframe • *Systems thinking:* Links transportation, land-use, and jobs • *Deliberative:* Attempts at citizen engagement with focus on neighborhood empowerment • *Reflexive:* Has been updated regularly; updates part of state mandate

Plan Purpose and Development Process

Why did the six cities studied here develop strategic plans? In Portland and Seattle, planning is required by state law. In Heidelberg and New York, strong mayors decided to develop a strategic plan to implement a future vision for the city. In Berlin, outside funding and grassroots engagement led the city to develop its Local Agenda 21. Hamburg's growth vision served as a political statement rather than a concrete plan. In short, these cities developed different sorts of plans for different purposes. This is supported by the range of plan development processes, engagement of key actor groups and the public, and policy sectors included. However, despite these differences, most of the plans studied here were developed for a similar reason: to address population growth. This goal tended to divide elites from the public in the cities studied, as decision-makers saw growth as either inevitable or desirable, while the public was more skeptical. There appears to be a link between a plan's conceptualization of growth and the level and type of public engagement, although this may be confounded by city size.

Growth was a central issue in plans, but handled in different ways. In New York, Mayor Bloomberg argued for environmental protection as a sound way to grow the city's economy. *PlaNYC*, that city's plan, emerged to some degree from the fusion of economic development and climate policy goals. Plans in Hamburg and Berlin focused on attracting skilled workers and investment. But even plans led by administrators were often based on or induced by population projections that suggested an influx of new residents in the medium term, as seen in all six cases.[24]

The differences between cities are subtle and seen mostly in the language used to address growth. It demonstrates whether these cities see growth as a goal to be pursued or an inevitability to be managed. Size appears to be a relevant factor. The largest cities (New York, Berlin, and Hamburg) welcomed expected population increase or even included marketing strategies to attract new residents, while the medium-size cities (Seattle and Portland) sought to accommodate growth in keeping with past trends. No particular differences between countries were seen.

However, beyond size, there is evidence that the type of public inclusion in planning affected the city's conceptualization of growth. Specifically, elite-driven processes embraced or sought growth and focused on development goals such as international competitiveness and growth of highly skilled migrants. This was most evident in Berlin's new urban development plan, as well as plans in Hamburg and New York. By contrast, plans developed with more input from the public are more inward-looking, seeking to promote neighborhood vitality, quality of life, and amenities for current residents. *The Portland Plan*, which was developed with the help of an open-ended questionnaire given to residents, used an equity framework as its core vision. In Heidelberg, changes made through the public comment process questioned the city's growth projections and called for a focus on affordable housing rather than accommodation of new residents. Also in Hamburg, outside groups questioned the city's growth vision, which was developed by the governing political coalition. At the same time, the most "pro-growth" cities were also the largest by population, suggesting this also had a role to play in how cities perceived growth.

Participation had additional effects related to plan implementation. Participation increased both public awareness and ownership of plan among those included, but

24 Such projections have been challenged as having a hidden normative component (Lloyd, 2000) and were criticized by citizens in several case cities.

could reduce ownership among those excluded from the development process. Berlin's approach to Local Agenda 21 allowed engaged citizens and stakeholders to co-design modules of the city's plan, but this may have reduced political ownership. In Heidelberg, the draft plan included a comment period where political parties, civil society groups, and industry organizations could suggest changes to the text, many of which were included in the final plan. New York's sustainability advisory board built support for the plan with the stakeholder groups included on the board, but created resentment among those excluded from the process. Seattle's devolution of aspects of implementation to the neighborhood level reduced opposition to the plan. This implies the possibility of a more modular approach in larger cities where city agencies focus more on facilitating neighborhood-level activities (such as in Seattle) or soliciting input from organized groups (as in Heidelberg). In addition, plan development processes often used merely superficial participation; in these cases, involved participants sometimes complained about a lack of voice. Here, more work is needed to develop processes that include diverse actors meaningfully in order to build support for the final plan.

Engagement of diverse policy sectors and actor groups also aided implementation. In New York, *PlaNYC* helped promote cooperation between city agencies as well as with private partners to achieve common goals, such as with the Million Trees initiative; in Heidelberg, the vision created in the urban development plan enjoyed broad public support. In addition, cities could increase coordination with higher levels of government. Seattle and Portland demonstrated the benefits of coordinating with higher levels of governments through plans and their requirements. However, additional opportunities remain. For example, goals listed in Germany's national sustainable development strategy, such as reducing greenspace conversion, could be better linked to state or city-level action.

However, inclusion alone was not sufficient. For example in Heidelberg, public interest in and awareness of sustainability appears to have faded over time. At the same time, broadly inclusionary processes take longer and require far more complex education and coordination (as seen in Seattle, Portland, and Heidelberg). The required time commitment was given as one reason why New York opted for a more expert-oriented stakeholder approach rather than broad-based public input.

Cities may benefit from further guidance on whom to include in such processes. For example, would congestion pricing have passed in New York if the city had included different actors in its sustainability advisory board or town hall meetings? It appears that public engagement and stakeholder-based approaches both provide benefits. It is important to design input processes in a way that smoothes implementation. Here, a more consensus-oriented political climate (such as that seen in Portland and Seattle) appears to help, but the factors that could improve this are little understood; although several people I talked to in Heidelberg mentioned a newly collaborative political climate as Beate Weber's most important contribution as mayor, none could describe specific actions that had created this climate. Beyond this, it is important to make sure that actors in charge of implementation are involved in development, as in many cases their exclusion led to lack of buy-in or even opposition to plan elements.

Simply developing a plan is not enough: in Berlin and Hamburg, plans were not used in policymaking after their development. Implementation requires high-level political support. Mayoral engagement in plan development and implementation was seen as critical in Heidelberg and New York. Plan implementation can be aided by planning requirements that institutionalize plans, protecting them from obsolescence after a change of political administration, as seen in Portland and Seattle, although planning mandates for both cities came after city actors lobbied state legislators to pass such requirements. In general, this suggests that political motivation is a prerequisite for effective planning. As such, strategic plans are a useful tool of urban sustainability governance only when political support exists. How to build such support remains an important question.

Plans and Sustainability Principles

Although all the cities studied here developed some sort of strategic plan, the last section highlighted that sustainability was generally not the main goal of these plans. Thus it is important to evaluate the extent to which the plans studied here incorporated sustainability principles. This is done on the basis of both the plans' content and their development process. Most of the plans studied here addressed all five principles to some extent, but to different degrees. Long-term considers and policy integration, which reflected complexity, were seen as the two principles best incorporated into plans and planning. These principles are also key to good planning, independent of sustainability consideration. Most plans included deliberative elements by including the public or key stakeholders in key ways, but more work is needed to help structure and guide such processes. Acknowledgment of limits and reflexivity were the least well incorporated principles, particularly considering the importance of growth in most plans. In many cases it was difficult to operationalize these principles for the planning context: for example, how long is "long-term"? As such, this analysis should be understood as an early attempt to evaluate urban strategy plans based on sustainability principles. Further development of such evaluative techniques is needed.

Acknowledgment of limits. This principle was the least well incorporated into the plans. To some extent, this reflects the challenge of operationalizing limits in an open system like a city. Some cities mentioned limits, but few policies were seen in these plans to reduce cities' resource use to some limited total. Still, climate change reduction targets were included in all plans, and some cities' land-use policies reflected an understanding of limited space. On the other hand, most of the plans studied here were developed to deal with growth. While Portland and Seattle sought to manage what they saw as inevitable growth, Hamburg's 2002 vision framed population growth as necessary for quality of life and, indeed, sustainability. Hamburg, New York, and Berlin discussed growth as both inevitable and positive for the city.

Long time scales. All plans studied used an initial timeframe of at least 20 years, suggesting limited relevance of this principle. Because much of a city's built environment must stand for 100 years or more, long-term planning can be advantageous. At the same time, projections have proved unreliable, making accurate long-term planning difficult. It is thus not necessarily advantageous for cities to develop plans longer than 20–30 years in advance. However, many plan updates did not extend the target year, reducing the long-term dimension of the plan. This issue also demonstrates limited reflexivity in planning processes.

Systems-thinking. All plans, by virtue of being strategic, acknowledged complexity or used a limited systems approach. New York in particular made explicit the interconnections between different policy areas; one of Hamburg's visions mentioned the complex interactions between policy sectors. Full inclusion of this principle would require coherent multi-level planning (that is, at the regional and state level), which several cities have attempted. The biggest shortcoming of the plans seen here is where they are developed and implemented within city government. If sustainability is truly to be a strategic issue, they city's plan would need to come from the mayor's office or a strategic position with jurisdiction over all city agencies. The extent to which planning departments have this position varied among these cities.

Deliberation. How this principle was incorporated varied widely between cities. Hamburg's vision documents appear to be developed exclusively by political actors; Berlin and New York used a stakeholder approach to develop *Berlin 2030* and *PlaNYC 2030*, respectively; and Heidelberg, Portland, and Seattle engaged the public in plan development. The way and extent to which the public is engaged appeared as well to affect plan purpose and content. Cities with more extensive public engagement processes focused more on quality of life issues for current residents, whereas plans with more elite involvement emphasized growth, particularly of highly qualified new residents. Inclusion of key stakeholders also increased broad ownership of plans. Here, it appears necessary to strike a balance between meaningful public engagement and stakeholder-based inclusion to reduce the number and criticisms of those left out of the plan development process.

Reflexivity. One key reflection of this principle was in comprehensive planning requirements in Germany as well as Oregon and Washington, which call for regular plan development and updates. Heidelberg's sustainability test (discussed more in Chapter 6) also attempted to include this principle, but lack of motivation in city agencies appears to reduce its effectiveness. Although several plans were updated and modified, new plans or visions sometimes failed to even acknowledge existing plans. This was seen most clearly in Berlin, where the team developing the current strategic plan did not refer to the city's Local Agenda 21, even as a failed case. It appears that full incorporation of this principle requires motivated actors, as reflexivity would aim to evaluate past efforts in order to make future planning more effective.

Conclusions

Strategic plans are vision documents that function above the concrete policy level. This makes them a useful governance instrument and well suited to urban sustainability governance. In addition to supporting several key findings from the literature, such as the usefulness of planning mandates in aiding implementation, the six cases studied here revealed two key lessons. First, the motivation of key actors was seen as critical for the plan's success, suggesting that plans are only useful in cases where political interest in sustainability exists. Second, the strategic plans analyzed here were generally created to deal with expected population growth, with sustainability as a method more than a vision. As such, sustainability principles were considered to different degrees.

First, the problems of implementation seen in several cases suggests that having a plan cannot substitute a lack of motivation of key actors to realize sustainability. Plans that do not resonate with local decision-makers, such as Berlin's Local Agenda 21, tend to have little impact after completion. At the same time, a plan that enjoys only the support of the current government, such as the vision documents seen in Hamburg, also plays a limited role in agenda-setting. A plan is a blueprint: creating it is only a first step. For a plan to be implemented, the main actors responsible for that implementation must accept the plan's vision. Especially in cases where the plan has no legal standing, buy-in of the public and key decision-making actors is critical to the plan's effectiveness. In cases that lacked a motivated lead actor—such as the mayor—or a legal mandate to push the plan into implementation stages, strategic plans served as a partisan vision of those who developed it, without seeing further use in the city. Therefore, although strategic plans can promote the sort of integrated decision-making critical to urban sustainability governance, they appear to function only in cases where such motivation exists.

In addition, sustainability is generally not the primary purpose of the plans studied here. Instead, sustainability is used to link different policy goal—such as economic development and climate policy—and set a more strategic course for decision-making. Instead, a primarily function seen throughout the cases was to deal with growth, particularly population growth. The plans generally focused on the need to build, renew, and adapt key infrastructure to invite or accommodate new residents and provide them with jobs. Although these goals were at times challenged by grassroots and advocacy groups, they were a key factor for plan development in all six cases. However, in cities with more deliberative processes that engaged the public, growth became seen as a challenge to current residents rather than a goal to purpose. In this way, the process of plan development had an important impact on how growth was framed.

Sustainability principles were seen to some degree in the plans studied here. In particular, the plans dealt with complexity and attempted some form of public or stakeholder deliberation. Acknowledgment of limits was seen less frequently, as were reflexive mechanisms, which appeared particularly lacking when plans were updated. Finally, most plans were developed with a time horizon of 20 years, a relatively long time scale for political decisions, but less so in terms of sustainability.

The cases did not vary widely across countries, but rather based on city size, political culture, and planning process. The key distinguishing factors seen here to affect how these principles were engaged were: political leadership, state planning requirements, civil society engagement, and extent to which the public was included in the process. These factors will also be seen in the next chapter, and will be explored further in the examination of sustainability-minded groups in Chapter 7.

Chapter 6
Quantifying Values with Sustainability Indicators

The last chapter highlighted the potential of strategic plans to help incorporate sustainability principles into policymaking in six U.S. and German cities. Sustainability indicators, like strategic plans, are a governance instrument considered critical for realization of urban sustainability. They can do so by playing different roles, such as operationalizing the concept of sustainability, measuring local policy effectiveness, and more generally building consensus and partnerships on sustainability issues. However, as with strategic plans, indicators often failed to live up to their promise (see Chapter 4). If indicators do not fulfill the roles suggested by the literature, what role do they actually play in urban sustainability governance? This chapter examines this question by asking: *Why do cities develop sustainability indicators?* and *how do they help incorporate sustainability principles into political decision-making?*

The analysis discusses the various roles sustainability indicators play in urban sustainability governance. However, the key lesson is that motivation is a determining factor for use. Actors do not use indicators that are not consistent with their interests. Although indicators may effect subtle values changes at the margins, they are not an effective tool at leading such an undertaking by themselves. The analysis presents three key ways indicators are used, based on who is using them: advocacy, strategic, and administrative. When indicators are developed or used with one of these purposes in mind, they can indeed fulfill their promise. In this way, indicators can serve to increase motivation of actors to engage in sustainability issues when used by interested advocates, as required for strategic plan success. Development must be linked to target users, however, for indicators to be successful.

Because each city uses sustainability indicators in a different way—and most cities have developed more than one indicator set—the cases vary in length. Within each section, up to four "cases" of indicator use are discussed, with the focus on describing the indicator set, how and by whom it was developed, and the main functions it serves in urban decision-making. Table 6.1 presents an overview of the indicator sets studied in each city. Three key types of uses emerge from the cases, which are discussed at the end of the chapter.

Table 6.1 Sustainability indicator sets discussed in six case cities

City	Indicator set(s) studied
Berlin	• Indicators for *BerlinStudie* • B-Nax • Local Agenda 21
Hamburg	• HEINZ
Heidelberg	• Sustainability test • Sustainability reports • FEST indicator report • Baden-Württemberg's *Leitfaden*
New York	• PlaNYC
Portland	• Portland Plan • Oregon Benchmarks
Seattle	• Sustainable Seattle indicators • Cascadia Scorecard

Berlin: Indicators, Re-invented and Re-ignored

Berlin's experience with sustainability indicators highlights a lack of interest in sustainability among the city's political actors and scattered but generally ineffective efforts by outside actors to use indicators to incorporate sustainability principles into decision-making. Although the city parliament has commissioned several reports and inquiry commissions related to sustainability, including passage of a Local Agenda 21 in 2006 (see Chapter 5), these efforts were almost universally ignored soon after publication. Although this may be related to other factors, such as political or administrative dysfunction or the city's relative lack of urgent ecological sustainability problems, the city's lack of institutionalized sustainability indicators does indicate that sustainability lacks political resonance as a framework for policy in Berlin.

The city's first set of sustainability indicators came from interested external actors and was called the Berlin Sustainability Index (*Berliner Nachhaltigkeitsindex*, or B-Nax). A large grant awarded to the Institute for Future Studies and Technology Assessment (*Institut für Zukunftsstudien und Technologiebewertung*, or IZT) by the local lottery foundation was distributed among numerous efforts working on sustainability governance, one of which to develop the B-Nax (BER01). The indicator set was a joint project of the *Fachforen der Berliner Agenda 21* (subject councils of the Berlin Agenda 21), the *Projektagentur Zukunftsfähiges Berlin* (Project Agency Sustainable Berlin), and the *Unabhängigen Institut für Umweltfragen* (Independent Institute for Environmental Issues, or UfU) (Berlin Senat, 2009, 47), which indicates a massively participatory indicator development process. The city's 10 *Fachforen* (subject councils) included 15–20 members from the administration, citizens' groups, environmental and social nonprofits, and other groups on issues related to Berlin's Agenda 21, such as education, mobility, and participation (BER01; Schmidthals and Kliche, 2004). Through strategic discussions on the city's goals in these thematic areas, the *Fachforen* suggested indicators with strategic goals for 2020, which became the basis of B-Nax's 60-indicator set (Schmidthals and Kliche, 2004; BER01).

Despite the massive effort developing the B-Nax, the indicator system was never explicitly linked to city policymaking. When the city passed its Local Agenda 21 in 2006 (see Chapter 6), it decided not to use the B-Nax, arguing that the indicators' thematic areas and approach were incompatible with the new plan (Berlin Senat, 2009, 48), although someone involved in developing the B-Nax told me that this was an excuse not to have to link the plan to performance measures (BER01). Instead, the 2009 report's 63 indicators focused on the three thematic areas of sustainability administrators warned the B-Nax team against using (Berlin Senat, 2009; BER01). A second report should be released in 2014, but no one I spoke with gave any indication that such a report was being developed. Indeed, even the B-Nax development team's suggestion that indicator data collection be institutionalized within the state's statistical agency was ignored, as this was considered by politicians to be too much expensive additional work (BER01).

The Berlin-Brandenburg state statistical agency did release a sustainability indicator report in 2012 that includes 16 core indicators and a number of additional subindicators (Amt für Statistik Berlin-Brandenburg, 2012). Although many of the new indicators can also be found in the B-Nax and/or the city's Agenda 21 report, some previous areas of interest—such as regional and international goals included in the B-Nax—are absent in the statistical agency's report. In addition, the foreword written by Berlin's planning and environment minister mentions neighborhood Local Agenda 21 efforts, but not the citywide document passed in 2006 (Amt für Statistik Berlin-Brandenburg, 2012, 4–5).[1] At the same time, he describes the indicator report as "a good foundation for the further development of indicators and goals in pursuit of a sustainable and integrated urban development" ("*eine gute Grundlage, auch für die Weiterentwicklung der Indikatoren und Ziele im Sinne einer nachhaltigen und integrativen Stadtentwicklung*," Amt für Statistik Berlin-Brandenburg, 2012, 5). Time will tell whether this holds true; the sustainability advocates I spoke with are not holding their breath (BER01, BER03, BER04).

As discussed in the previous chapter, Berlin's Local Agenda 21 has not been used to guide policymaking. In addition, although the city is home to numerous groups interested in sustainability in general and in Berlin specifically, their work is generally more focused on projects such as urban gardening or the re-municipalization of the city's electricity infrastructure (see Chapter 7). As such, Berlin has seen the least use of sustainability indicators in policymaking of the six cities studied here. Indicators have not been used by administrators, political actors, or advocacy groups. In addition, the efforts to develop indicators seen here have started from scratch, declining to use previous indicator sets to

---

1 This is consistent with his public comments during development of the city's new strategic plan (see Chapter 5).

which developers feel no connection. The main disconnect, it appears is between developers and users: B-Nax was developed by interested advocates for use by policymakers. It remains to be seen whether the statistical agency's indicator set meets the same fate.

Hamburg: Using Indicators to Contest Political Visions

In contrast to Berlin, Hamburg has seen a diverse and active use of sustainability indicators in its local policy debate. Hamburg's experience with sustainability indicators highlights how contested sustainability can be in the city, as well as the key role played by two actors in setting the terms of debate: the city's planning and environmental agency (*Behörde für Stadtentwicklung und Umwelt*, or BSU) and the *Zukunftsrat* (Future Council), a diverse network of actors that promotes sustainability in the city. In Hamburg, sustainability indicators have been used as a political tool to shape the city's definition of sustainability and spur it to action, both with the *Zukunftsrat*'s HEINZ indicator set and its critical response to government-developed growth indicators. This section describes the role of the HEINZ indicators in contesting city policy and advancing the *Zukunftsrat*'s definition of sustainability in the city.

Hamburg has a long-standing and well-used sustainability indicator set called HEINZ (*Hamburger Entwicklungs-INdikatoren Zukunftsfähigkeit*, or Hamburg's development indicators of sustainability), developed by the *Zukunftsrat*, a network of over 100 member groups focused on sustainability in the city (see Chapter 7). HEINZ primarily serves as an advocacy tool to raise awareness and frame the issue of sustainability (Döring and Pahl, 2000, 89, cited in Teichert, 2000, 167), but the indicators have also been used in an official capacity, primarily by the city's planning and environment agency (*Behörde für Stadtentwicklung und Umwelt*, or BSU). HEINZ has risen in prominence within and outside the city since its inception, aided by the assistance of BSU staff and financial resources, and become a fixture of the *Zukunftsrat*'s work on sustainability.

The *Zukunftsrat* was created by the city government to serve as the extra-governmental organ responsible for implementation of sustainability in Hamburg. In 1998, it created a working group for indicators, which included BSU staff and indicator experts from the local university (HH02, HH03). It should be noted that, because the group's work is almost exclusively conducted on a voluntary basis, the group's indicators were developed due to the interest and efforts of a group of interested parties (HH02; Schindler, 2011). The working group released its first indicator report in 1999, a report card (*Zeugnis*) that graded the city's sustainability using several key indicators. This report card received an overwhelming public response, partly due to the government coalition at the time of the relatively environmentally friendly SPD and Green parties. It was presented and discussed at the city's interagency sustainability working group, several city agencies responded in writing to it and the city and state education agencies ordered several hundred copies. In all, about 80 requests for the report card came from across Germany

(Geffers, 2000, 68). In addition, the report card became a major focus of public outreach, being posted in subway stations and serving as a key component of the *Zukunftsrat*'s public information campaign, which included 40 events, an Agenda newsletter, and a brochure on the indicators. The report card was distributed and promoted the indicator reports with city agencies, politicians, associations, and at sustainability conferences (Geffers, 2000).

In 2003, the *Zukunftsrat* released the HEINZ sustainability indicators, which were similar to the original sustainability indicator set. HEINZ has become a major part of the group's work, being published almost annually since 2003. The group sees these indicators as a way to promote dialogue between actors, gain traction on sustainability issues in decision-making, and encourage public identification with the city (Menzel, 2008; HH02).

Over time, the HEINZ indicators have generally shown negative social trends and positive economic trends. The 2005 report included 32 goals for 2020 that were explicitly normative, such as an unemployment rate below 3 percent, 1 percent inflation, and environmental programs in 10 percent of Hamburg businesses. Measures based on GDP were removed beginning with the *Zukunftrat*'s 2005 indicator set. However, it should be reiterated that the group's role as a pressure group meant that these goals were not linked to specific policies or even an official municipal strategy, but served to present the organization's policy vision for a sustainable Hamburg. At the same time, former *Zukunftrat* speaker Menzel (2004) highlights the lack of follow-up on indicators by a city with an officially proclaimed interest in sustainability. Although Hamburg committed itself to developing sustainability indicators when it signed the Aalborg Charter, "there has been no effort from parliament or city hall to take up this task" (Menzel, 2004, 63). Indeed, although the city's good performance on its own indicators—economic growth, eco-certificates—contrasts with the more negative picture presented by the *Zukunftsrat*.

In 2005, the city of Hamburg released *Monitor Wachsender Stadt* (Monitor for the Growing City), its own catalog of indicators to measure progress on the city's strategic vision, *Metropolis Hamburg* (see previous chapter). In response, the *Zukunftsrat* took a more adversarial approach with its indicators. Opposition from the *Zukunftsrat* and other corners, such as the publication in 2004 of a volume that critically questioned Hamburg's growth vision (Altrock and Schubert, 2004) led the city to include HEINZ indicators to measure its strategic goal of "securing quality of life and sustainability of the city" in its updated set in 2005 (Menzel, 2008; Schindler, 2011, 206).[2] In this way, city decision-makers could easily tack a readily available sustainability indicator set onto its own list of progress indicators. However, the compromise of an add-on meant that the government was more interested in neutralizing criticism from the *Zukunftsrat* and others but had not re-considered its overarching goal of quantitative economic and population growth (Menzel, 2008).

2 It should be noted that although these are both academic publications, each was written by a speaker of the *Zukunftsrat*.

HEINZ has seen a wide range of use in recent years. The indicators played a role in party politics: the SPD and Green parties used HEINZ as ammunition against the governing coalition in the 2005 Bundestag election, highlighting poor performing indicators as evidence that CDU policies were ineffective (Menzel, 2008, 168). They have raised the interest of the city government as a subject of Senate debate in 2004 and 2007, when the *Zukunftsrat* explained its approach and compared it to the *Monitor* (Menzel, 2008, 169). They have been used by the city's sustainability office within BSU and were instrumental in the city's successful application for European environmental capital in 2011 (HH01). The indicator set is also seen as a model throughout Germany, despite its lack of incorporation into official decision-making in Hamburg (for example, Schmidthals and Kliche, 2004).

At the same time, HEINZ indicators have served primarily as an advocacy tool. Döring and Pahl (2000) state:

> In terms of general requirements for a reporting system with the goal of sustainability, the *Zukunftsrat*'s reporting provides little support for political decision-making processes as well as for selection and evaluation of sustainability-related political measures (political orientation and evaluation functions). It is much more an instrument to provide information to the politically interested public. (89; cited on p. 167 of Teichert, 2000)

In other words, HEINZ is not intended to be used by political decision-makers, but to influence more public debate on sustainability. This suggests an attempt by the *Zukunftsrat* to increase interest and motivation in sustainability in city decision-making through more general advocacy. It remains to be seen whether sustainability advocates can break through to get their values better represented in city decision-making processes, but Sustainable Seattle's similar experience detailed later in this chapter highlights the effects such indicator sets can have.

Hamburg's experience with sustainability indicators is complicated and mirrors the city's ambivalence related to sustainability. The city has seen active, ongoing development and use of sustainability indicators in the form of HEINZ, whose indicators have been collected for over a decade. At the same time, the HEINZ indicators were developed by an outside group and have been used almost exclusively as an advocacy tool with little linkage to decision-making. As such, HEINZ indicators appear to be targeted at creating interest in sustainability in Hamburg. In addition, although the city has not yet developed strategic policies related to sustainability, employees at the urban development and environment ministry have worked closely with the *Zukunftsrat*, as members of the indicator working group, as providers of data for the indicators, and as the group's main source of funding (HH01; HH02). In this way, HEINZ indicators serve a similar role to Sustainable Seattle's indicator reports in the 1990s.

Heidelberg: Administrating with Indicators

Heidelberg has perhaps the most institutionalized use of sustainability indicators within local administrative and political decision-making processes, and thus an experience vastly different from Hamburg's and Berlin's. Through the city's sustainability test (*Nachhaltigkeitsprüfung*), virtually all decisions made by the city council (*Gemeinderat*) are checked against the sustainability goals listed in the city's urban development plan (STEP, see previous chapter). In addition, the city has released a series of indicator reports linked to the STEP and been supported to some extent by Baden-Württemberg's state-level sustainability strategy. However, a lack of coordination between the city government and FEST, a research institute that has specialized in sustainability indicator development and developed an early indicator set for the city, presents an interesting missed opportunity in Heidelberg. These three examples of indicator sets and their use in Heidelberg highlight a lack of coordination with organizations beyond city agencies and limited political interest in operationalizing sustainability in the city.

As part of its implementation strategy for the city's urban development plan (see Chapter 5), the city of Heidelberg passed a quality-control protocol throughout the city's agencies in the form of a sustainability test. The sustainability test, which has been in place since 2005, requires that all proposed resolutions to the city council include a discussion of what goals set forth in the STEP are promoted as well as explain if any goals are harmed by the proposal (Amt für Stadtentwicklung und Statistik, 2009).

The test has two sections. First, it calls the person preparing the resolution—generally a city administrator—to list the STEP goals that are affected, state whether the effects are positive or negative, and provide a brief explanation. The second part calls for a brief discussion of any conflicts between goals and any other important considerations. In this way, administrators and city council members are constantly reminded of the strategic goals laid out in the STEP.

A proposal signed by mayor Würzner describes the purpose of the sustainability test:

> This required test includes a thematic examination of the STEP goals and is conducted by the leading agency. Supervision is provided by department heads. Through this, the most important statements on goals affected by the action are listed and explained. When conflicts between goals exist, these should be critically discussed (*Drucksache* 0199/2009/IV, 2009, 3.1).

The city reported on the effects of the sustainability test in 2005, finding three generally positive effects: 1) the issue of sustainability and STEP goals became more present in the daily deliberations of city administration and the city council; 2) interagency perspective and evaluation were strengthened to some extent; and 3) the explanations provided by the sustainability test have increased transparency and promoted reflection on policy activities and goals (*Drucksache* 0170/2005/IV, 2005, 3.2).

The city planning agency released a second evaluation of the sustainability test in 2009 (Amt für Stadtentwicklung und Statistik, 2009). This brief report discusses how often different STEP goals and thematic areas, as well as neighborhoods, were most often addressed in 2007 and 2008. Of the 338 decisions that included a sustainability test in 2007, goal conflicts were mentioned only 14 times, most frequently between issues of balanced budget, climate and pollution prevention, and greenspace preservation (Amt für Stadtentwicklung und Statistik, 2009, 3). In 2008, 341 resolutions included the sustainability test; only 11 included goal conflicts, with greenspace preservation as the most frequently cited (ibid., p. 6). A further check by the city's Agenda office, responsible for implementation of the STEP as the city's Local Agenda 21, found that at least 97 percent of sustainability tests were properly filled out in 2008 and 2009 (*Drucksache* 0199/2009/IV, 2009, 3.2). As part of this report, the city reduced the scope of proposals requiring the sustainability test as well as its length in order to "further optimize" the process and "reduce its time requirements" (ibid.). The test demonstrates a way indicators connected to the city's strategic vision can be used to make decision-making more reflexive toward sustainability.

Heidelberg's sustainability test has recently been incorporated into statewide planning in Baden-Württemberg (HD01, HD02), another indication of its efficacy. At the same time, this tool may be suffering from implementation fatigue. One administrator in Heidelberg expressed worry that the sustainability check is only given cursory consideration by over-worked employees (HD03). Those performing the check often simply copy and paste information from previous documents and omit potential conflicts between goals (HD03). At the same time, it was agreed that the sustainability test does serve to incorporate sustainability concerns into decision-making in some way and keep it part of public debate (HD01, HD03). It thus appears to be a useful tool for the incorporation of sustainability—as measured by the city's strategic plan—into decision-making processes.

In addition to linking sustainability to municipal decision-making through the sustainability test, Heidelberg has released a series of reports (in 2004, 2007, and 2011) using indicators to evaluate implementation of the city's urban development plan, the STEP (*Stadt Heidelberg*, 2011).[3] These reports include indicators linked to plan goals and include changes in individual indicators over time, an individual rating in the form of a stoplight, and a general rating for each of the 10 thematic areas of the plan.

The STEP indicator reports serve primarily to evaluate implementation of the city's strategic plan. Like the city's sustainability test, they function to link the strategic plan to specific measures, coordinate action, and increase accountability. They are thus an important reporting mechanism for progress in realizing the city's vision.

3 The city's first report, in 2002, did not include indicators (City of Heidelberg, 2002).

The final indicator set discussed here was developed by an external group, the Protestant Institute for Interdisciplinary Research (*Forschungsstätte der Evangelischen Studiengemeinschaft e.V.* or FEST). FEST is a major developer of indicator sets in Germany, which has worked on projects at the municipal, state, regional, and national levels. It has been instrumental in the state-level indicator set used by Baden-Württemberg, but surprisingly has not played a meaningful role in Heidelberg's sustainability reporting.

FEST released an early sustainability indicator report for Heidelberg in 1997 (Diefenbacher et al., 1997). The research institute developed the idea to create a sustainability indicator report for the city after two of its researchers taught a class at the University of Heidelberg in 1994–1995. After securing letters of intent from the city and the Rhein-Neckar regional governments (in 1995) and funding from the state environment agency (in 1996), FEST developed its indicator report (HD04, HD05). Because the indicators were developed externally, they appear to have had problems connecting with decision-makers. When the first STEP indicators report was released in 2004, it did not link to FEST's sustainability report for Heidelberg, which was not further pursued.

In the late 1999, FEST developed a common municipal indicator set to be implemented statewide in a research project commissioned by four German states, including Baden-Württemberg (Diefenbacher et al., 2009). The research team developed an indicator framework (*Leitfaden*) through discussions with municipal representatives from the participating states, interested agency officials from other states, and indicator experts. The indicators were then tested by several municipalities to determine their practical value as well as the availability of necessary data. The final framework has gone through four iterations (the fourth edition was published in 2009 and a fifth was planned for release in fall 2013, Diefenbacher et al., 2009; HD02, HD05) and forms the foundation for municipal sustainability indicators in Baden-Württemberg. Over the past decade, FEST has done extensive consulting with municipalities to incorporate this framework into their planning and develop additional indicators to measure issues of importance to them. The framework report also includes information on which government agency should be responsible for collecting data on each indicator and recommend indicator reports be published every two or three years.

FEST's state-level indicator set has seen traction in municipalities other than Heidelberg, which benefit from data and resources provided by Baden-Württemberg (HD02). This indicator set also provides a strategic framework to help cities frame decisions and report progress, making it a tool of multi-level governance. However, coordination across agencies appears to be a problem, as the state statistical agency recently released its own sustainability indicator report based on a different framework and composed of incompatible indicators (Diefenbacher et al, 2009; HD02; Statistisches Landesamt Baden-Württemberg, 2012).

It may seem odd to include a description of FEST's sustainability indicator framework in a discussion of sustainability indicators in Heidelberg, because the city does not use this framework in its sustainability reporting. Indeed, the city's indicator set, which is linked to its urban development plan, is thus separate from the state-supported approach. This is not necessarily a problem, as state-level sustainability staff prefer to give cities such as Heidelberg, which are leading in terms of sustainability, autonomy to implement sustainability as they wish (HD02).

To conclude, Heidelberg has done perhaps the most work of the cities studied here to use sustainability indicators as part of urban sustainability governance. Through its sustainability test and sustainability reporting, the city asks agency staff and city council members to consider sustainability, as linked to the city's urban development plan, in most important policy decisions. In this way the city has developed a strategic policymaking process in which individual policies are tied to strategic goals, with sustainability used as the glue that binds them together. Heidelberg's approach, unique among the cities studied here, represents a primarily administrative use of indicators. They have been so solidly incorporated into city decision-making that there is little discussion of the extent to which they measure sustainability principles, or Heidelberg's most pressing sustainability goals. This suggests a lack of subsequent debate regarding sustainability at the strategic level, indicating a possible lack of reflexivity regarding the city's sustainability goals. Such debate could be re-opened by a strategic plan update process.

At the same time, the city has not taken advantage of opportunities to collaborate with other indicator developers, such as the state of Baden-Württemberg and FEST. Such lack of coordination could be interpreted in several ways: that indicator development is more appropriately conducted by users than external experts, that city administrators do not require additional resources or expertise, that these different actors simply have not found ways to collaborate, or that city actors prefer not to work together for whatever reason.

New York: Indicators for Strategic Performance

Like Heidelberg, New York uses performance measures tied to its sustainability plan as an important way to incorporate sustainability into decision-making. Unlike Heidelberg's plan, however, PlaNYC does not cover all city decision-making, instead focusing on the built environment and a few other issues (see Chapter 5). For this reason, the plan serves as one of many major city initiatives, rather than an umbrella under most city policy falls. The performance measures included in PlaNYC are also the only use of sustainability indicators I found in the city.

Performance measures are a central feature of New York's sustainability plan, as they focus the initiatives on causes that the city government can impact and provide concrete, measurable goals that can be evaluated by the public. Their function is not so much to monitor key aspects of sustainability in the city, but to measure how well it has implemented the specific initiatives laid out in the plan. They also seek to make PlaNYC—and sustainability—more accessible to New Yorkers, by "embedding this stuff into the fabric of people's lives"

(NY02), especially important in a city that seems to lack public momentum for sustainability. In addition, the plan's goals are easy to track, include medium-term goals in addition to 2030 targets, and form a key part of its annual progress reports. Indeed, they can likely be best understood as performance measures, as their key purpose is to track implementation of PlaNYC. As such, the indicators used by the city focus on areas targeted by PlaNYC initiatives rather than general information about sustainability-related issues in the city.

A typical example is the MillionTrees initiative, which has its goal built into the name: the city aims to plant one million new trees by 2017 (City of New York, 2007). This simplifies program evaluation and has led to increased collaboration between city agencies as well as with external partners (NY03). At the same time, a discussion with one city forester revealed an unintended consequence of the initiative: although it has led to a dramatic increase in the parks department's budget, much of this has been earmarked for the purchase and planting of new trees, precluding its use from maintaining existing trees (NY03). Because of this, the city may end up spending more money to maintain the same tree cover.

New York's experience with indicators is, therefore, fairly unique among the cases studied. Like Heidelberg, indicators serve a primarily administrative function, although several people I spoke with hoped that the measures linked to PlaNYC's initiatives would draw public attention and support for the plan. In contrast to cities such as Hamburg and Seattle, sustainability has little public resonance in New York, meaning there is little discussion or use of PlaNYC's performance measures outside of its progress reports. Although Mayor Bloomberg has discussed progress made on the plan's initiatives, sustainability indicators in New York serve more of an administrative purpose: setting common goals, integrating efforts between different city agencies and partner organizations, and measuring progress and effectiveness.

Portland: Mismatched Indicator Developers and Users

Portland has seen less work on explicitly termed sustainability indicator sets in recent years than some other cities studied here. This may be due to the higher incorporation of sustainability into strategic indicators developed as early as the 1980s. What appears more likely than lack of interest in sustainability is its realization through more holistic strategic initiatives such as the city's strategic plan. This section examines the indicators in the *Portland Plan* as well as progress measures by the statewide Oregon Progress Board.

The Portland Plan, the city's strategic plan, includes 12 "measures of success," general areas of interest to measure implementation. Each citywide measure, such as "equity and inclusion," "growing business," and "complete neighborhoods" is linked to more concrete indicators: for example, "income distribution" and "diversity index" (City of Portland, 2012, 105 & 108). Because the *Portland Plan* is not an explicit "sustainability" plan, these indicators are better understood as performance measures than sustainability indicators, much like the indicators used in New York's *PlaNYC*. Many of the indicators include goals for 2035, while others (such as the diversity index) appear to play merely a descriptive role.

Beyond the vision of social equity suggested by the goals (the first indicator is income distribution) is the innovative way in which these goals are measured. For example, the fourth measure of success is "prosperous households," which is measured not in GDP per capita or household (as is typical), but "percent of economically self-sufficient households," as measured by households meeting the threshold set by the Self-Sufficiency Index, a refined metric that equals about twice the poverty line (City of Portland, 2012, 113). Qualitative measures such as resident satisfaction are also included. However, economic measures such as export production rank appear inconsistent with the principle of limits—as realized by closed systems—and thus worthy of further consideration. Still, sustainability appears to be embedded in the fabric of many of the indicators, even if it is not explicitly stated.

Due to state-mandated comprehensive planning (see Chapter 5), the measures described in the *Portland Plan* are linked to specific policies and by law must be tied to policymaking in the city. The broad public visioning process that fed into plan development gives the indicators in the *Portland Plan* a solidly deliberative underpinning reflective of local values. Although use of the term "sustainability" faded during the course of the development process, sustainability principles of environmental quality, social equity, and consensus-based comprehensive planning are reflected in the plan. This implies that although the term sustainability plays a less important role in policymaking in Portland than other cities studied here, its principles have been incorporated into political decision-making.

The second indicator set discussed here is the Oregon Benchmarks, a statewide set of indicators created by the Oregon Progress Board. Created in 1989, the Oregon Progress Board was an initiative of Governor (and former Portland mayor) Goldschmidt that developed a massive set of indicators created to measure progress on the state's strategic plan *Oregon Shines*. The Oregon Benchmarks have been cited in the literature as a best practice in terms of good governance, strategic planning, and sustainability in the last two decades (for example, Barg, Anielski, and Waddell, 2006; Köckler, 2005; PCSD, 1996). They earned the state an Innovations in Government award by the Ford Foundation and Harvard University's Kennedy School of Government (Kissler et al., 1998). The 1991 package of benchmarks approved by the state legislature included 158 indicators in three categories—"exceptional people, outstanding quality of life, and diverse, robust economy"—with 17 lead benchmarks linked to key problems identified in the strategic plan (Tryens, 2004, 3).

The Progress Board was chaired by the governor and consisted of 14[4]—later reduced to nine—members, citizen leaders chosen to "[reflect] the state's social,

4 This number includes two ex officio members, a youth representative and the director of the state Department of Administrative Services (Tryens, 2004, 3), which is why

ethnic and political diversity" (Oregon state website, 2007, in Misaras, 2007).[5] It had "little statutory authority" to call on state agencies to realize the goals included in the benchmarks, instead drawing on the institutional relationships of board members and considering implementation "the responsibility of all Oregonians" (Tryens, 2004, 2).

Governor John Kitzhaber later created a 46-member task force to update the strategic plan, which was released in 1996. At this time, the state legislature "overwhelmingly" passed legislation to continue the Progress Board's work, specifically calling on the Board to "encourage the discussion and understanding of critical global and national trends," "develop 'a strategy that addresses the economic, social, cultural, environmental and other needs and aspirations of the people of Oregon,'" and develop goals with measurable indicators to be reported on every two years (Kissler et al., 1998, 356).

The Oregon Benchmarks were intended to provide politically relevant information to state-level decision-makers in order to inform policymaking. However, Misaras (2007) found that despite politicians' recognition of the indicators' potential value, they were not actually used in shaping policy. Her findings generally supported other writing on policy use of indicators in finding that indicators are generally used to support existing narratives (see Chapter 4).

Then-state representative (and later losing Portland mayoral candidate) Jefferson Smith argued in 2011 that the state benchmarks were inadequately incorporated into decision-making processes. He provided three suggestions for change: "having all incoming state legislators go through an orientation process to familiarize them with the benchmarks and thereby help them start thinking about policy ideas that could address problems identified by the indicators, having the benchmarks be linked to the governor's state of the state address[,] … [and] having a required impact statement that would show how the proposed legislation would influence benchmarks" (Kroh, 2011, 5). However, Smith's attempt to better institutionalize the Oregon Benchmarks was at odds with state Republicans, who considered them partisan and moved to defund the project when they gained control of the legislature in 2009. The state's two-decade experiment with a comprehensive, deliberatively developed indicator set has thus come to an end.

The Oregon Progress Board served as a model in Portland, leading Mayor Vera Katz to create the Portland Multnomah Progress Board to help implement Portland Future Focus, the city's strategic vision in the late 1980s (Portland city auditor "history," 2008; VisionPDX, 2008, 5). A joint initiative between the city of Portland and Multnomah County, the board released a series of reports on

some accounts describe the Board as having 12 members (for example, Oregon Progress Board website in Misaras, 2007).

5 All information related to the Oregon Progress Board has been purged from the state government's website. Much of the account here is from several full-length government texts related to Board included as appendices in Misaras (2007).

implementation of the city's strategic plan in the 1990s and 2000s. However, this group has also been disbanded; its last report appears to have been published in 2009 (Portland city auditor's website, 2013).[6]

Both the state of Oregon and the city of Portland were pioneers in the area of performance measurement linked to public strategic plans. Portland has developed indicators to measure progress on its comprehensive plan that align well with sustainability principles. In addition, state-level requirements for plan implementation make it likelier that these indicators will be connected to policymaking at the city level in the future. As implementation began in 2013, it is too early to tell the full impact these measures will have on the decision-making process, but the Portland case does show that sustainability can be incorporated even into indicator sets that do not explicitly bear its name.

At the same time, more comprehensive city and statewide indicator efforts appear to have ended when their respective progress boards were disbanded in the late 2000s. The reasons appear to include partisan fears that the benchmarks would lead policymaking into an untenable "liberal" direction, as well as admission that such broad indicator sets do not actually see targeted use in policymaking. Both of these efforts are emblematic of the high expectations of sustainability indicators in the literature: that creating comprehensive sets of values we think are important can improve decision-making and help incorporate sustainability principles. However, the experience of the Oregon Benchmarks and the Portland Multnomah Progress Board confirm previous findings that indicators are not used as developers or researchers expect (Innes, 1990; Rosenström, 2006; see also Chapter 4).

This presents an important lesson relevant to sustainability indicators: because indicators used for policymaking are normative, they do indeed require a certain foundation of values that cannot be taken as a given and can be challenged by those who disagree with them. Indicator sets that do not meet the needs of expected users are not used or continued.[7] This highlights the importance of including users in indicator development as well as the necessary ongoing work to communicate indicators' function to the public and inspire more general ideational change.

Seattle: A Complex Ecosystem of Indicators

Seattle has a long history with sustainability indicators and has made extensive use of them in several areas to highlight the diverse range of their use (Pettibone, 2012).

6 From the City Auditor's website, "Note to user: The Portland Multnomah Progress Board is no longer in existence. These are static files and will not be updated" (http://efiles.portlandoregon.gov/webdrawer.dll/webdrawer/rec/3660820/, retrieved April 4, 2014).

7 Of course, it is also important to remember the specific context of partisanship in the United States, especially within the Republican party. Republicans in Clackamas County, Oregon, who feared the international governance implications of sustainability and ICLEI voted to revoke the county's membership in that organization.

Two initiatives in particular—Sustainable Seattle's indicators in the 1990s and the Sightline Institute's Cascadia Scorecard—show how an easy acceptance of sustainability by political and administrative decision-makers as well as grassroots and environmental organizations can speed incorporation of indicators into city decision-making processes.

The first iteration of indicators in Seattle has been referred to as the first set of sustainability indicators in the world (Holden, 2007). Led by the volunteer group Sustainable Seattle in 1993, indicators were developed by a participatory process that included over 150 participants—elected officials, nonprofit and institutional leaders, and engaged citizens—who created measures that generated enthusiasm for sustainability from residents but were seen by some as lacking methodological rigor (Brugmann, 1997; Holden, 2006; Mitra, 2003). The goal of the project was to create measures for values important to the community, such as "children living in poverty" and "public participation in the arts" (Sustainable Seattle, 1998). The group sought

> to initiate and contribute to a city-wide conversation about sustainable development that in time would lead to a new system of urban policy and practice. To S2 [Sustainable Seattle] members, it made sense to design their project and its process by the key components of this new system: an emphasis on measuring and monitoring, collaboration between citizens, experts and decision-makers, linking indicators into a holistic framework, and the sharing of decision-making power via better information, communication and dialogue (Holden 2006). In taking this approach, S2 worked toward an impact on the policy process that did not take the form of specific new policies. (Holden, 2007, 254)

The group's audience was primarily community members and the media, with government and business actors considered secondary (Mitra, 2003, 22). Its main goal could thus be understood as the development of a conversation on sustainability and impact the policy process beyond specific policy proposals.

Whether through either amazing foresight or sheer luck, the group's first indicator report was realized during the height of development of Seattle's comprehensive plan, *Toward a Sustainable Seattle* (see Chapter 5). However, rather than focusing exclusively on a public awareness campaign as in Hamburg, Sustainable Seattle tried to offer the report to the city government (SEA05). In Mitra's (2003) account, the group's volunteer basis made it impossible to a finance large-scale publicity campaign (23), which may have led the group to make this decision. The city planning department decided to essentially cover the report's printing costs by buying the first 2,000 copies to distribute at planning meetings, as its chief planner fear, "once it belonged to the city, then the city would mess around with it to make sure that the city looked better" (SEA05). This kept the indicators independent—allowing them to be critical of city policy if necessary—but brought them informally into the planning process.

Sustainable Seattle released a total of four indicator reports during the 1990s, but struggled with enthusiasm and funding over the next several years

(Holden, 2006). The organization's membership and its mission have changed over time (Holden, 2006; SEA06). This could be seen as a testament to the quick acceptance of its projects by city government—some neighborhoods appropriated the indicators for mandatory reporting required by the state's *Growth Management Act*—or the different interests of new members. At the same time, several founding members of the organization now hold important positions in city decision-making: Richard Conlin was elected to the city council and Alan Durning founded the Sightline Institute (originally Northwest Environment Watch), which seeks to incorporate sustainability issues into public debate in the Pacific Northwest (see Chapter 7).

Some studies have expressed disappointment with the lack of longevity of Sustainable Seattle's indicator set (Holden, 2007; Mitra, 2003). However, I have argued elsewhere that the group's evolution is evidence of the ecosystem-like behavior of well-functioning governance systems (Pettibone, 2012). According to that line of reasoning, Sustainable Seattle had to change its mission because it was essentially successful in meeting its goal: introducing sustainability into public discourse. The group recently spun off a project on happiness indicators that uses original survey data to measure Seattle residents' happiness. The Happiness Project has received positive feedback from the city council, a main projected user of the data; as of 2012, the project was working on a more scientific follow-up survey (SEA07). Sustainable Seattle now concentrates on inspiring more concern for sustainability within the business community, which it sees as an important area of focus (SEA06).

In addition to causing change within the city, the work of Sustainable Seattle has spurred numerous indicator projects throughout the world. Sustainable Seattle's work influenced at least nine other sustainability indicator sets within the region over the next decade (Holden, 2007). These ranged from indicators related to government plans at the state, county, and state level to the Sightline Institute's Cascadia Scorecard, which measures the entire Pacific Northwest region and was influenced by Sustainable Seattle's indicators (Holden, 2007). Although it is unclear whether this leadership has led the region to become more sustainable or the liberal attitudes of its residents caused the indicators to be developed in the first place, Seattle is setting the bar for other cities and regions. In addition, members of Sustainable Seattle were invited to at least 22 conferences around the world in the 1990s and its indicators won an award from the United Nations (Holden, 2006). The group's indicators have served as a best practice example in several German research projects (Heiland et al., 2003; Lang, 2003; Wittek et al., 2002),[8] giving them broad international resonance.

8 However, misrepresentations in some research on these indicators may have created a false impression of indicator development and use that has precluded research into the role of external groups in urban sustainability governance. For example, Lang (2003) characterizes Sustainable Seattle's indicator set as created by the city in cooperation with other groups (180–81), which misrepresents its development and role in Seattle.

Another project led by a nonprofit organization is the Cascadia Scorecard, a regional sustainability indicator set developed in 2004 by the Sightline Institute. The scorecard tracks seven key regional trends—energy, wildlife, sprawl, population, economy, health, and pollution—and includes distance-to-target ratings. In addition, some indicators are measured in innovative ways: the "economy" score comes from an original composite index of "incomes of typical households, the unemployment rate, the share of northwesterners living in poverty, and the share of children living in poverty" (Northwest Environment Watch, 2004, 19). Creation and continued use of such statistics may have affected, for example, Portland's selection of economic security measures in its recent comprehensive plan.

Although still published annually, the purpose and presentation of the scorecard have changed since its inception, suggesting lessons learned by Sightline in understanding its usefulness. The indicators were originally released annually in report form as a sustainability scorecard. However, the current focus is more on analyzing the data and reporting trends:

> What people often are looking for with data is, they're looking for ways of telling their story. And sometimes, the same piece of data can be important in multiple different stories, in opposing storylines. So simply having 'the data' [is not enough]: facts are slippery things. (SEA08)

In other words, the Cascadia Scorecard is no longer seen as a straightforward reporting mechanism, but is used to shape the debate related to sustainability in the region. This is part of Sightline's work on sustainability in the region, which is discussed in greater detail in Chapter 7. These two indicator projects in Seattle highlight the role that sustainability indicators can play in terms of advocacy. The changes in how these indicator sets are used—seen in the end of Sustainable Seattle's indicator project and evolution of the Sightline Institute's Cascadia Scorecard—demonstrate the nuanced approach to indicator use and the active role played by advocates in seeking more general values change through indicators. Sustainable Seattle's work is also interesting when compared to the *Zukunftsrat*'s HEINZ indicator set in Hamburg. The divergent experiences of these groups underscores the advocacy role that indicators play when developed by outside groups. Specifically, whereas Sustainable Seattle appears to have been successful in getting sustainability incorporated into decision-making (as seen by its inclusion in the city's strategic plan), the *Zukunftsrat* has still not been able to get sustainability onto the political agenda. For this reason, Sustainable Seattle has moved on to other work, whereas the *Zukunftsrat* continues to use its indicators and other actions to increase politicians' interest in sustainability. More generally, the key role played by these and other outside groups demonstrates the importance of further research on these sustainability-minded groups, which are the focus of the next chapter.

Analysis

The six cities examined reveal a rich variety of ways that sustainability indicators are used in political decision-making. Cities use sustainability indicators for numerous reasons (as seen in Table 6.2), which can be distilled to essentially three broad purposes: to advocate for change, to advance strategic action, and to measure political performance.[9] At the same time, several examples were seen here of indicator sets that saw no use. As such external development of indicators is pervasive in the literature, this serves as an important lesson that indicator development must be linked to expected users in order to have political value.

The different experiences of these six cities present a rich story of indicator use. Politicians, administrators, and sustainability-minded groups were all seen to use indicators, albeit in different ways. Lack of use appeared to be due to a disconnect between indicator developers and users: it was seen in cases where indicators were developed by one group for use by another. This suggests the limited role external advisers can play in using indicators to inject sustainability directly into decision-making. However, some outside groups have developed a different, more successful strategy: the advocacy role played by externally developed indicators that civil society groups themselves use highlights a different way for outsiders to use indicators in urban sustainability governance. Although concrete use of performance measures was less evident, it is clear that such indicators aid policy implementation and reporting.

9 Although it is not part of this research, I saw some evidence of a fourth use of indicators, individual indicators or standards. Here, individual indicators such as such as green building codes, fair-trade food labels, or walkability scores seek to link key attributes (energy efficiency, organic agriculture, or walkable neighborhoods) to sustainability and further define how to measure these attributes. However, their focus on individual rather than policy change puts them beyond the scope of this discussion.

Table 6.2 Sustainability indicator sets and their uses

City	Indicator set(s)	Developed by	Main users	Functions
Berlin	Indicators for *BerlinStudie**	Citizen advisory groups	None	• [Issue framing] • [Decision support]
	B-Nax*	Research institute Subject advisory councils	None	• [Issue framing] • [Decision support]
Hamburg	HEINZ	Advocacy group	Advocacy group, politicians	• Issue framing
Heidelberg	Sustainability test	City govt	City govt	• Decision support • Policy integration
	Sustainability reports	City govt	City govt	• Performance measurement
	FEST indicator report*	Research institute	None	• [Issue framing] • [Decision support]
	Baden-Württemberg	State govt, research institute	State govt, local govts	• Performance measurement Policy integration
New York	PlaNYC	City govt	City govt, project partners	• Performance measurement Policy integration
Portland	Portland Plan	City govt/ public	City govt	• Decision support Performance measurement
	Oregon Benchmarks*	Govt-organized advisory board	State govt	• Decision support Performance measurement
Seattle	Sustainable Seattle indicators*	Advocacy group	Advocacy group, city govt	• Issue framing
	Cascadia Scorecard	Think tank	Think tank	• Issue framing

* Indicates indicator set no longer in use. Brackets [] indicate expected use.

The following sections discuss the three broad uses of indicators in more detail, focusing on how they contribute to urban sustainability governance in the six case cities. They draw on experiences in the six case cities and highlight specific indicator uses by key actor groups.

Indicators for Advocacy

Interest in sustainability cannot be taken as a given in urban decision-making processes. For this reason, early change agents often agitate for its inclusion in political decision-making. Indicators developed by outside groups were often used by the groups themselves for advocacy purposes. Outside groups in several of the cities examined here used sustainability indicators for advocacy purposes: both to bring sustainability concerns into the public debate and to link sustainability to important local issues. Early volunteers at Sustainable Seattle developed an indicator set that city planners used to build momentum for a discussion of sustainability in developing the city's strategic plan. The region's Cascadia Scorecard developed and maintained by the Sightline Institute aims to present a vision of a sustainable Cascadia bioregion and draws attention to negative trends. However, the change of emphasis in both of these projects suggests that such an advocacy use of indicators can be fulfilled as sustainability begins to be taken seriously by politicians. That the HEINZ indicator set developed by the *Zukunftsrat* in Hamburg is still being updated annually reflects the lack of traction of the group's vision of sustainability in city decision-making there. At the same time, Seattle's experience reinforces the notion that advocates use indicators to call for change: the Sightline Institute's more ad hoc use of its Cascadia Scorecard is based on the need to tell a coherent story of sustainability.

This advocacy use of indicators could be seen as a first step to call on city decision-makers to incorporate sustainability into policymaking processes. Indicators used for advocacy:

- Seek to reframe issues into a sustainability narrative;
- Agitate for political change or increased attention for an issue;
- Require high comprehensibility and political salience among the target audience;
- Are less methodologically rigorous.

If sustainability indicators are primarily used for advocacy in the city, this may be a sign that sustainability principles have not been incorporated into decision-making: advocates are agitating for that to happen. This model of indicator use can thus help to identify whether sustainability truly is part of a city's urban governance. The findings from the case studies suggest that advocacy is an important early stage for indicator use and by extension incorporation of sustainability principles into decision-making processes. Although this use was seen in both Seattle and Hamburg, the diverging experiences in these cities suggest how advocacy can incorporate sustainability principles into political decision-making. When successful, advocates may move on to other issues.

This has implications for research on both sustainability indicators and urban sustainability governance. First, sustainability indicator research has often focused on explicit use of indicators in political decision-making. The introduction of an advocacy function for indicators highlights an important "pre-policymaking" role that is insufficiently understood. Indeed, several case studies where authors have expressed disappointment with the lack of use by political decision-makers may have missed the important work being done in this stage. Therefore, more work should be done to understand this function of indicator use and what factors lead to a progression to the next stages of the model, strategic and institutionalized use.

Strategic Indicators

Politicians did not appear to use sustainability indicator sets in a systematic way, but did refer to specific indicators to tell political stories. They also developed indicator sets as part of a strategic vision and for administrative use to evaluate implementation of strategic policy instruments such as comprehensive plans. The performance measures in PlaNYC are an excellent example of this: they served to both distill the plan's numerous policy initiatives into easily digestible numbers (such as the Million Trees initiative) and allowed monitoring of performance on policies attached to the plan. The sustainability test in Heidelberg sought to connect a strategic definition of sustainability with indicators that aimed to directly affect policy decisions. Although the skepticism of some involved in the process suggests that the test is not completely successful, it was the most comprehensive attempt to use sustainability indicators to steer policymaking, and is thus worthy of further study. In addition, the FEST indicator set used by Baden-Württemberg presented municipalities with the state's strategic vision for sustainability. In short, administrators were seen to use indicators to support decision-making and measure performance.

In this stage, sustainability principles may have begun to be incorporated into decision-making, as decision-makers themselves develop sustainability goals and indicators to measure them. Strategic indicators are:

- Developed by government;
- Political, but enjoy some level of consensus among decision-makers;
- Usually linked to issues under government jurisdiction.

Here, sustainability has entered the "mainstream": it has been discussed to some extent in the public square and is understood as an important goal of policymaking. At the same time, strategic indicators may only cover a small part of all political decisions made in the city. Such indicators are also more contested then those in the next stage: although there is some level of consensus, it may only be within the current governing party or even a single powerful mayor (as seen in New York). For this reason, sustainability should not be seen as fully institutionalized in city decision-making processes, but as having gained a toehold.

Indicators as Performance Measures

The third use of indicators could be seen as the holy grail among indicator researchers. Here, the importance of sustainability is so entrenched as to be taken as a given. Indicators are institutionalized: data is collected regularly by reputable, politically neutral statistical organizations. They may not be actively used, but frame how local problems are understood and form an informational foundation in many political decisions. In this stage, sustainability is part of the decision-making framework. To some extent, this overlaps with strategic use. The key difference is that here indicators guide daily administrative decision-making, rather than targeted—and often temporary—high-level political initiatives.

This use was seen in cities with a more institutionalized strategic plan, such as Heidelberg, New York, and Portland with the *Portland Plan*. At the same time, early indicator sets in Berlin saw no use and the Oregon Benchmarks was discontinued for partisan reasons. These experiences suggest again that sustainability should not be taken for granted as a framework for policy. Even in Heidelberg, where sustainability indicators form the basis for a sustainability test for political decision-making, administrators worry that the test has become too automated to allow critical reflection of policy proposals' sustainability implications.

Sustainability indicators that fulfill this role:

- Show trends to provide background information for decisions
- Need to have long-term data availability
- Create general understanding of issues, which can subtly guide decision-making (for example, commute times vs. miles of highway)

This role for indicators has been an important part of the literature on indicators to date (for example, Gehrlein, 2004; Heiland et al, 2003; Innes, 1990; Misaras, 2007; Pastille, 2002).

How Indicators Contribute to Urban Sustainability Governance

The last three sections presented the three main uses of sustainability indicators in the six case cities. This model suggests a simple linear progression, but that should not be expected in all cases. Some cases saw strategic use of indicators without progressing through an advocacy stage (as in Heidelberg and New York): in both cases, the key "advocate" for sustainability was the mayor. In addition, indicators developed for strategic or information purposes without buy-in from users (such as B-Nax in Berlin and the Oregon Benchmarks) can lack the support necessary to be effective, especially when developed exclusive by external groups for use by others. The model therefore does not necessarily reflect a progression that naturally occurs in all cases, but one may lead to more institutionalized incorporation of sustainability into policymaking. At the same time, this conceptualization suggests an increasing institutionalization of sustainability in the policymaking context, an issue worthy of further exploration, as it could aid the evaluation of urban sustainability governance in a more systematic way.

Table 6.3 presents the typology of indicator use developed from the case studies. This understanding differs from a large portion of the literature on sustainability indicators, much of which has been dedicated to developing indicator sets (for example Diefenbacher et al, 1997; Häberli and Schneider, 2002; Mori and Christodoulou, 2011; Sawicki, 2002; Trockel, 2013), which this research showed were not used. In general, the literature has focused more on indicator development and use within government, excluding what this research has found to be a key use by outside groups for advocacy.

Table 6.3 Different uses of sustainability indicators

	Advocacy	Strategic	Performance measurement
Purpose	• Issue framing	• Policy integration • Performance measurement	• Decision support • Performance measurement
Characteristics	• Agitate for political change or increased attention for an issue • Require high comprehensibility and political salience among target audience • Are less methodologically rigorous	• Developed by government • Political, but enjoy some level of consensus among decision-makers • Should be related to issues that government can change	• Show trends to provide background information for decisions • Need to have long-term data availability • Create general understanding of issues, can subtly guide decision-making
Key users	• Sustainability-minded civil society groups	• Politicians	• Administrators

Conclusions

This chapter discussed why six cities in the United States and Germany use sustainability indicators and how they contribute to urban sustainability governance. Sustainability indicators appear to play three key roles: advocacy for change, presenting a strategic vision, and measuring performance. Different actor groups were seen as taking the lead in performing each role, which were conceptualized as part of an evolving process towards incorporating sustainability principles into political decision-making. First, civil society groups developed indicator sets with high political relevance in order to raise awareness about sustainability and advocate for its inclusion in decision-making. If successful, politicians developed and used indicators to create a strategic political vision of sustainability for the city. This vision was implemented at the administrative level with the help of indicators to measure performance. In this way, each actor used indicators at a different stage of institutionalization to advance sustainability. Indicator sets were created and abandoned as needed; in general, the trend appears to be towards more ad-hoc use of indicators except as performance measures. In addition, indicators need to be developed in conjunction with anticipated use: those sets developed by one group for use by another were not pursued further in any of the cities studied. Finally, this research highlighted the active engagement of civil society groups in development and use of indicators, which has not yet been explored in the literature. As such, this issue will be pursued further in the next chapter.

Chapter 7
Building Motivation: Sustainability-Minded Groups

This study uses a conceptualization of urban sustainability governance that focuses on the incorporation of sustainability principles into political decision-making processes. The last two chapters presented the role played by two governance instruments—strategic plans and sustainability indicators—in achieving this. Although both instruments *can* promote urban sustainability governance, this does not mean that they actually *do*. Instead, their effectiveness is based on developers' and users' understanding of and interest in sustainability. It is therefore possible to develop and use plans and indicators that have no relation to sustainability principles. At the same time, plans and indicators developed with careful attention to sustainability principles may not be used if actors involved in implementation or use are not motivated by these principles. In short, although both strategic plans and sustainability indicators can aid the transition to urban sustainability governance, they require actors to be motivated by sustainability concerns in order to succeed. Such motivation can be seen as a prerequisite for the effectiveness of sustainability governance instruments.

The case study analysis in the previous two chapters highlighted several organizations that appeared to play an outsized role in incorporating sustainability principles into political decision-making processes. In Seattle, Sustainable Seattle led the city's sustainability indicator efforts; several of its founding members now hold high-level positions in Seattle politics. In Hamburg, the *Zukunftsrat* challenged the city's urban development vision and has released sustainability indicators for nearly a decade. Clearly, these organizations play an important role in sustainability governance. However, such groups have not yet been the focus of a concerted research agenda (Holden's work on Sustainable Seattle is a noteworthy exception).

The definition of sustainability-minded groups used here emerged inductively from the case study analysis on strategic plans and sustainability indicators. Sustainability-minded groups are defined here as organizations that seek to define sustainability and engage in the transition to sustainability at the local level. They may be linked to government structures, as Germany's Council for Sustainable Development at the national level, or be completely independent, such as the U.S. Green Building Council or BUND. Their defining feature, however, is their core mission: to advance sustainability within some geographically or politically defined area. Many sustainability-minded groups can be considered think tanks, as they conduct research and advocacy work on sustainability issues. (Indeed, the Sightline Institute, based in Seattle, calls itself "a think-and-do tank.") But others advocate for sustainability through project work, such as composting or urban gardening. This chapter seeks to highlight the importance of studying sustainability-minded groups in understanding urban sustainability governance. It presents some first steps in conceptualizing what sustainability-minded groups are and the role they play in motivating action toward urban sustainability governance. It addresses the question: How are sustainability advocates, in the form of sustainability-minded groups, advancing sustainability principles in political decision-making processes?

Outside Sustainability Groups in Six U.S. and German Cities

This section examines the civil society groups working to define and advocate for sustainability in the six case cities in the United States and Germany. Some cities—especially Berlin and New York—have numerous smaller groups that could not all be profiled here. The purpose is therefore not to provide an exhaustive list of all groups engaged in sustainability activity in these cities, but to provide an illustrative list and discuss the most active groups in more detail. Although the analysis here is exploratory, I hope to demonstrate the importance of these groups in urban sustainability governance and thus the value of further research into their work.

Berlin: Grassroots Groups and Think Tanks

Berlin has a diverse and active community engaged on sustainability issues in the city as well as nationally and internationally. The city has one main organization focused on sustainability in Berlin—Berlin 21. In addition, a variety of other groups focus primarily on two additional types of work: implementation of local projects and academic research related to sustainability at the national and international level.[1] This section focuses on the work performed by Berlin 21, local sustainability initiatives, and Berlin-based think tanks to highlight the variety of ways in which these groups contribute to sustainability governance generally as well as the lack of explicit connection to local-level policymaking.

Berlin 21 is a civil society organization created by the Berlin Senate to help guide the city's Local Agenda 21. However, since then, it has slowly decreased its financial support for the organization, finally ending it in 2013 (BER03). The organization is the most active on sustainability issues within the city, but has stopped seeking to influence political debates in order to focus on areas where

1 Beyond these groups are businesses that use a sustainability framework, such as the e-mail service provider Posteo and sustainable clothing designers and retailers including Kollateralschaden and Supermarché, respectively, which are beyond the scope of this research.

it can have greater efficacy. As such, the group hosted a series of themed events on aspects of sustainability such as education, diversity, and transportation in 2013 and in 2014 published a brochure that profiles 41 initiatives working on important aspects of sustainability in the city (Berlin 21, 2014). This brochure seeks to highlight the concrete work done by grassroots groups in order to connect them and improve their effectiveness. At the same time, Berlin 21's events hope to inspire local interest in sustainability, although these often see the same small group of participants (BER03).

In addition to Berlin 21, the Grüne Liga is a network of environmental groups active in Berlin that also organizes the Network21Congress (*Netzwerk21Kongress*), Germany's annual conference of Local Agenda 21 and sustainability coordinators (BER04). This work has given the group connections and funding for projects such as urban gardening, educational activities, and demonstrations. The Grüne Liga helped in developing the city's Local Agenda 21, but also functions largely as an environmental group in line with BUND and NABU.

Finally, Berlin has seen growth in think tanks and research institutes dedicated to sustainability in recent years. Organizations such as Ecologic and the Independent Institute for Environmental Issues (UfU) conduct research for clients on energy, climate, and sustainability issues. As seen in Chapter 6, UfU was instrumental in developing the B-Nax, a sustainability indicator set ignored by decision-makers. Currently, the group works on projects from the local to European level; staff members are also engaged as volunteers in groups like Berlin 21 (BER01). In addition, the relatively young Institute for Advanced Sustainability Studies (IASS) was founded by Klaus Töpfer, a high-ranking politician who served terms as environment and urban planning minister and as executive director of the United Nations Environment Programme (UNEP). This organization has so far been successful in securing funding for numerous research projects, although it is less active in Berlin (and based in nearby Potsdam).[2]

Berlin is home to numerous groups dedicated to sustainability issues in the city and beyond. However, these groups have not had traction in changing political priorities and as such have retreated primarily to project work. In addition, funding concerns present a major problem for organizations like Berlin 21.

2 Berlin and nearby Potsdam have become a hub of sustainability, climate change, and environmental research, with several research institutes based at the city's three public universities as well as independent research organizations like IASS. However, these institutes do only limited work related to sustainability in the city itself.

Table 7.1 Outside sustainability groups in Berlin

Organization	Purpose
Berlin 21	• Previously coordinated Local Agenda 21; after falling out with administrators, funding was eliminated • Connects local initiatives, projects related to sustainability in Berlin • Holds public events on issues related to sustainability
Grüne Liga Berlin	• Environmental network that conducts projects on sustainability in urban context • Coordinates annual Netzwerk21 meeting of local sustainability practitioners in Germany; helped coordinate development of city's Local Agenda 21
UfU, IASS, ecologic, etc.	• Academically oriented think tanks conduct research on sustainability, often funded by grants or project funds

Hamburg: Stepping on the Senate's Toes

More than the other two German cities studied here, Hamburg has seen the active involvement of a sustainability-minded group—the *Zukunftsrat*—in trying to motivate political action on sustainability in the city. This network organization has worked for over a decade to frame sustainability in Hamburg with its sustainability indicator set HEINZ (discussed in Chapter 6) and its critical stance on the city's urban vision. As such, the *Zukunftsrat* is an excellent example of a sustainability-minded group.

The *Zukunftsrat Hamburg* (Future Council Hamburg) is the main organization devoted to sustainability in Hamburg. Created by the Hamburg Senate in 1997 to help coordinate civil society Agenda 21 activities, the *Zukunftsrat* was originally intended as an extra-governmental counterpart to the city's environmental agency (Geffers, 2000). However, political developments since then have meant that the *Zukunftsrat* is often the main actor interested in promoting sustainability in Hamburg (HH02). The group is an open network of various non-governmental institutions in the city, from the Chamber of Commerce to environmental organizations to HafenCity GmbH, although it is the coordinating committee that forms the organization's "beating heart" (HH02). Founded with 30 members (Menzel, 2008), the *Zukunftsrat* today has about 100 member institutions, although they vary widely in terms of engagement within the network.

The city provided the organization enough annual funding to pay for a part-time administrative position and a small office (HH07). This has meant that most of its work—developing reports and indicators, holding public meetings, communicating sustainability issues with the public—is done on a volunteer basis.

Because of this, most of the group's initiatives are developed on an ad-hoc basis, based on the interest and time commitments of its volunteers (HH02).

The *Zukunftsrat*'s longest running initiative is the HEINZ sustainability indicator set, which is updated annually (see Chapter 6). In addition, the organization holds several large public meetings each year on aspects of sustainability, such as education, art, and indicators of well-being. These functions seek to promote dialogue, raise aware of sustainability, and connect the term to local policy issues. The *Zukunftsrat* is a key advocate for sustainability in Hamburg.

Core members of the group have also released or contributed to several key advocacy and academic publications. These reports seek to define sustainability in the city, particularly by challenging the political visions that have emerged from recent governments and presenting an alternative vision more rooted in the principles of sustainability. Three in particular stand out as important: those by GERMANWATCH (2000), Altrock and Schubert (2004), and BUND Hamburg, Diakonie Hamburg, and Zukunftsrat Hamburg (2010).[3]

In 2000, the Hamburg regional group of GERMANWATCH published a book on Local Agenda 21 in Hamburg, which provided several critical voices on sustainability problems in the city, from a proposal for health care for immigrant prostitutes, criticism of military exports, a proposed to reform bond provision for ecologically damaging projects, and calls for debt forgiveness in developing countries. *Zukunftsrat* speaker Menzel's (2000) essay focuses on the city's understanding of sustainability, which he criticizes as "more preamble and projects than program and process."

In 2004, a group of civil society actors and academics published an edited a volume devoted to examining cities' desire for growth (Altrock and Schubert, 2004). Altrock and Schubert's (2004) edited volume on the meaning of urban growth visions combines contributions on Hamburg with experiences from other cities around the world in order to more deeply examine the theme of urban growth. In addition to the defense of growth by the mayor at the time (von Beust, 2004; see Chapter 5), the volume includes a chapter by the speaker of the *Zukunftsrat* criticizing the city's growth vision and calling for sustainability to be used as a guiding principle in its place (Menzel, 2004).

The third major publication, a joint project of BUND Hamburg, Diakonie Hamburg and Zukunftsrat Hamburg (2010), is the most recent attempt to create interest in sustainability in Hamburg. This report is in some ways an update of the 2000 GERMANWATCH publication and includes a number of concrete policy proposals with suggestions on how to get them passed and implemented. The editors are extremely critical of Hamburg's political actors: they argue that the

3 In addition to these works are research projects on sustainability in Hamburg with an external audience. Speaker Delia Schindler completed a dissertation on urban sustainability governance in Hamburg (Schindler, 2011) while working on the NEDS project, which focused on ways to operationalize the contested nature of sustainability and used Hamburg as a case study (Bauriedl, Schindler, and Winkler, 2008; Luks, 2007).

2002 growth vision was Hamburg's final refutation of sustainability as a guiding principle (BUND Hamburg et al., 2010, 10). Although there have been noteworthy projects, the city's indicator set shows its real priorities: economic and population growth and improving the city's international reputation.

All of these publications demonstrate active attempts to motivate the public and key decision-makers to care about sustainability concerns. Still, although these books and reports provide an invaluable time capsule to researchers of sustainability in the city, local actors are skeptical of the effect they have had on decision-making in Hamburg and frustrated at the lack of high-level political support for sustainability (HH02; HH03). It is also difficult to devote the energy necessary to the issue on an almost exclusively volunteer basis. Although its original role was envisioned as an equal partner with the city's environmental agency in implementing sustainability in the city, the *Zukunftsrat* has taken on a more adversarial role since its creation. Interview partners from both city administration and the *Zukunftsrat* understand the organization's role as "stepping on the Senate's toes" (HH01; HH02).

The *Zukunftsrat*, with its diverse membership including business, environmental, and administrative actors, has the makings of a strong coalition for the implementation of sustainability in Hamburg. The council has sought to develop a stronger baseline understanding of sustainability issues in the city through regular publication of its sustainability indicator set. That it has achieved so much despite a small dedicated funding source highlights interesting possibilities for sustainability advocates in other cities. This experience is similar in some ways to that in Berlin, where advocates unsuccessful in shaping the political debate have shifted their focus to neighborhood projects and grassroots activities. This raises the question of how external institutions can develop dedicated funding sources as well as anchor integration of sustainability into political decision-making. One opportunity is presented by the funding strategy used in Heidelberg; Portland and Seattle present a second funding model.

Table 7.2 Outside sustainability groups in Hamburg

Organization	Purpose
Zukunftsrat Hamburg	• coalition of over 100 member organizations working to advance sustainability in Hamburg • has coordinated or contributed to numerous publications on sustainability in Hamburg to challenge political definition, particularly growth orientation

Heidelberg: The Number-crunchers

Sustainability has become an important umbrella term in Heidelberg. At the same time, the constellation of organizations working on sustainability in the city is somewhat different from the other cities studied here. Although the city has several local groups working on aspects of sustainability, from environmental and natural habitat protection to human rights and "real world economics," these groups have not united under the banner of sustainability or advocated in a major way for action related to sustainability in Heidelberg itself. However, Heidelberg is also home to two major research institutions, IFEU and FEST, that have helped the city and the state of Baden-Württemberg with numerous projects related to sustainability. This section thus discusses the work of IFEU and FEST as well as several groups in the city's new *WeltHaus*, an office space recently opened to build a network of like-minded groups on issues related to environment and development issues.

The Institute for Energy and Environment Research (*Institut für Energie und Umweltforschung*, or IFEU) is a nonprofit research institute started by scientists at the University of Heidelberg in 1978. IFEU has conducted countless research and consulting projects in the city, such as analysis of the ability of the city's urban development plan to serve as its Local Agenda 21 (IFEU and Stadt Heidelberg, 1997; see also Chapter 5). IFEU has also developed several tools and indicators to help measure different aspects of sustainability, in the areas of mobility, waste management, and more (HD07). Currently, IFEU is coordinating the city's efforts on education for sustainable development (HD04). The organization has established a close working relationship with the city, but its work is focused more on quantifying aspects of sustainability, such as comparing bus to U-Bahn emissions (Kutzner event, 2013), rather than developing or advocating for a particular sustainability vision. As such, the institute can be best considered something of a consulting firm for sustainability. The group's numbers-driven approach could be argued to promote a primarily administrative sustainability based on efficiency and technological change rather than a morally based call for social change.

In addition to IFEU, Heidelberg is home to the Protestant Institute for Interdisciplinary Research (*Forschungsstätte der Evangelischen Studiengemeinschaft e.V.* or FEST), which has done perhaps the most work on sustainability indicators in Germany, from coordinating the *Leitfaden* used by Baden-Württemberg and other German states to consulting on the country's national indicator set and helping to develop an alternative well-being index to replace GDP (Diefenbacher and Zieschank, 2011; Diefenbacher, Teichert, and Wilhelmy, 2009). In 1997, FEST developed a sustainability indicator set for the city of Heidelberg (Diefenbacher et al., 1997); however, as discussed in the previous chapter, their indicators were not used by city decision-makers. Interestingly, although the organization's indicators team has worked for two decades on sustainability indicator development and implementation, it has done relatively little work with the city of Heidelberg. People I spoke to suggested this was due to waning interest in sustainability in the city as well as the time and personnel resources necessary to maintain a sustainability indicator set (HD04; HD05). At the same time, Heidelberg has released regular indicator reports for over a decade. It may thus be possible that city administrators feel that they have the strategic points of sustainability management under control and thus do not need to call in FEST's expertise.

In 2012, the city of Heidelberg and several nonprofit organizations arranged an agreement to develop the *WeltHaus* (World House). Some smaller groups approached BUND Heidelberg with an idea for a joint workspace; BUND pitched the idea to the Gemeinderat and got €50,000 to renovate the office space in an old building near the main train station (HD06). Roughly a dozen groups moved into the space in October 2012, and are continuing to deepen their cooperation: they have an intranet, an e-mail list, and a speakers' council (HD06). The organizations housed in the *WeltHaus* came together with a common goal developed at a joint retreat: the transformation society. Specifically, "How do we effect a transformation that can put the world on a sustainable path?" However, as of summer 2013, the organizations still had not developed a joint Welthaus project (HD06).

Although the advocate I spoke with said that Heidelberg has numerous initiatives related to sustainability—"we're spoiled"—the lackluster list of projects seems to indicate the opposite (HD06). Several groups have successfully realized projects in the areas of fair trade, sustainable consumption, and habitat protection, but there is no organization like the *Zukunftsrat* that focuses on sustainability on a strategic or political level, by seeking to define the term within the city and get the public, politicians, and businesses on board. In Heidelberg, although there seem to be a significant number of projects that support sustainability, there is not really a watchdog group that makes sure the projects are on the right track.

At the same time, the city has established a funding strategy for key environmental groups that appears to be effective. Every year the city budget includes €50,000 for BUND to provide free environmental consulting for citizens, which helps pay for the group's overhead but keeps it other independent (HD06). This funding provides a baseline for the group's work.

The lack of grassroots engagement on local sustainability may be partly due to Heidelberg's size: with 150,000 inhabitants, the city has about one-quarter the population of Seattle or Portland. Another factor may be the active role that politicians and the administration have played in incorporating sustainability into local decision-making. I had the hardest time in Heidelberg of the six cities studied getting potential interview partners to speak to me. Numerous people responded that they did not know about or work on sustainability issues, even those in environmental or degrowth organizations and city administrators in the planning office. Sustainability does not appear to be a motivating factor for civil society actors in the city. Instead, the main definition of sustainability in Heidelberg appears to be in its decade-old urban development plan and the sustainability test used in administrative decision-making (see Chapters 5 and 6). Whatever the reason, the lack of sustainability-minded groups in the city may be one factor that has led sustainability to be "administered, not lived": motivation for sustainability is not constantly refreshed by interested civil society actors.

Table 7.3 Outside sustainability groups in Heidelberg

Organization	Purpose
FEST	• Christian-oriented research institute with decades of engagement on sustainability issues, particularly sustainability indicators • has conducted numerous long-standing sustainability indicator projects from municipal to national level, albeit almost exclusively outside Heidelberg
IFEU	• environmental research institute that has developed numerous targeted sustainability indicators and technical information related to sustainability from municipal to national level • often works with city government on sustainability issues, such as facilitating comprehensive plan development
BUND Heidelberg	• provide energy and environmental advice funded by city government • work on environmental issues in Heidelberg

New York: Mainstreaming Environmental Justice

The constellation of groups working on sustainability-related issues in New York is different from most of the other cases studied here. In contrast to cities like Berlin, Hamburg, and Seattle, New York has no major nonprofit institutions working to define sustainability within the city. Instead, the most important alternate frame related to the city's sustainability policies comes from the city's environmental justice (EJ) movement, several leaders of which served on the city's sustainability advisory board or within the city administration (see Chapter 5). In addition, numerous neighborhood groups have used the language of sustainability—including the term itself. Their work, however, has generally focused on concrete projects. This section discusses the work of both environmental justice and neighborhood sustainability groups in defining sustainability in the city.

Although environmental justice is not the focus of this book, several authors have related it to sustainability (Agyeman and Evans, 2003; Keil, 2007; Pearsall and Pierce, 2010). In brief, environmental justice relates to the unequal distribution of environmental problems, particularly the disproportionate siting of Locally Unwanted Land Uses (LULUs) in poor, often African American neighborhoods (Rosan, 2012). Rosan (2012) examines the extent to which environmental justice concerns are reflected in *PlaNYC*, New York's sustainability plan. She finds that the plan does a good job of incorporating EJ concerns: Many areas ignored in the original report have been taken up in updates or additional policies, such as the municipal solid waste plan, a residential green building program, and comprehensive green stormwater prevention policies. However, new problems have emerged, such as the rezoning of EJ neighborhoods for more residents even as wealthier white neighborhoods have been downzoned, which raises questions of equity in the city's efforts to grow by one million residents.

Inclusion of two EJ activists on the sustainability advisory board, as well as the hiring of New York City Environmental Justice Alliance (NYC EJA) director Eddie Bautista to work for the city in Mayor Bloomberg's second term, was considered "notable and historic" by some—but not all—in the EJ community, as their concerns had been traditionally ignored by the city (Rosan, 2012). Still, others, such as Miquela Craytor of Sustainable South Bronx (SSBX), complained that "'we weren't asked to the table'" and that "their suggestions for how to link PlaNYC to economic development and promote green jobs through green infrastructure did not make it into the plan, which represented a missed opportunity" (Rosan, 2012, 966). Rosan concludes that the city has begun to deal with EJ concerns by including activists, consulting on large-scale planning projects, and siting waste transfer stations in all five boroughs. At the same time, the city continues to pursue policies at odds with these efforts, such as siting a jail in the Bronx without public consultation. So "the result is that the same EJ and sustainability groups that often work closely with the City find themselves at odds with the City" (Rosan, 2012, 973).

Rosan's case study suggests that the inclusion of environmental justice advocates in the sustainability advisory board that developed *PlaNYC* as well as important posts in city government has better aligned the city's definition of sustainability with EJ priorities. My research supports this finding; several interview partners highlighted the importance of having EJ advocates at the table in influencing the city's plans for solid-waste transfer stations (NY01; NY05). Thus, the environmental justice discourse has been connected to sustainability in New York's political decision-making.[4]

In addition to the city's strong environmental justice movement, New York has seen a proliferation of borough- and neighborhood-level sustainability groups in recent years. Many—such as Sustainable South Bronx and Sustainable Flatbush—have sustainability in their names. Others include groups such as 596 Acres and the city's various urban and rooftop gardening groups. These organizations focus on concrete neighborhood projects, such as developing composting programs, planting organic gardens, and taking over vacant lots. They are less engaged in citywide or discursive activities, although this may be at least partly due to exclusion from official channels. At the same time, the inclusion of environmental justice groups on the city's sustainability advisory board suggests the need to build better coalitions between these organizations so

4 The election of Bill DeBlasio as mayor in November 2014 suggests that EJ is on the rise. DeBlasio's planned update to PlaNYC in early 2015 will add equity to the plan framework (MyNYC.gov e-mail newsletter, March 25, 2015).

that selected representatives can better reflect the needs and views of the larger group. One volunteer I spoke with said that her organization did not have the time or capacity to meet with representatives from groups in other boroughs, and indicated that such networking was more time-consuming than productive, saying, "We've got to clean up our own home" (NY06).

New York is home to some individual sustainability advocates, such as the people who wrote for *Sustainability Watch*, a series of articles that appeared in the *Gotham Gazette* and later through the Hunter College website (for example, Angotti, 2008; Angotti and Checker, 2011; Cowett, 2008; Katz and Baron, 2011). These pieces, also presented at the 2007 event held by the Institute for Urban Design that also became bound as *New York 2030 Notebook*, an edited volume composed of views on the city's new strategic plan, cast a critical light on how sustainability was being realized politically. However, the most recent *Sustainability Watch* piece was published in 2011, despite later interest in resurrecting the initiative (Hunter College, n.d.; NY10). At the same time, particularly flattering portraits of New York's plan by ICLEI USA and the Climate Desk environmental reporting initiative have likely reduced the reach of such critics (Bagley and Gallucci, 2013; ICLEI and City of New York, 2010).

Fragmentation among neighborhood groups and project initiatives was also seen in other cities, such as Berlin and Seattle. However, unlike those cities, New York does not appear to have an umbrella organization dedicated to defining sustainability at the city level. Whether this contributes to or is a product of sustainability's lack of resonance in New York is unclear. The lack of explicitly sustainability-minded groups working at the discursive level has allowed the city government and its partners to define sustainability essentially unilaterally. This has meant relatively little discussion or debate on what sustainability means for such a big city (Byles and Kazi, 2008, being the key exception) or how it is to be achieved in the city beyond Mayor Bloomberg's primarily development-based vision. The lack of a sustainability-minded institution working on advocating for sustainability in the city's discourse also reflects and maintains the limited traction the term has had in city decision-making.

Table 7.4 Outside sustainability groups in New York

Organization	Purpose
New York City Environmental Justice Alliance (NYC EJA), UPROSE	• Environmental justice organizations committed to highlighting and correcting historically unjust distribution of environmental problems in New York neighborhoods • Group representatives included in development and implementation of New York's sustainability plan
Sustainable South Bronx, Sustainable Flatbush	• Neighborhood groups focused on concrete projects linked to sustainability, such as organic gardening and compost collection
Sustainability Watch	• Published critical articles and reports on New York's sustainability plan

Portland: Sustainability By Any Other Name

As seen in the last two chapters, Portland developed important planning policies now linked to sustainability before the term became popular. Oregon, where Portland is based, has been a leader in state-level planning guided by growth management (see Chapter 5). Key advocacy groups have arisen to preserve Portland's planning culture and inform development of its Urban Growth Boundary (UGB), notably the umbrella group 1000 Friends of Oregon.[5] Formed as a coalition of rural and urban actors, 1000 Friends was founded by former governor and growth management proponent Tom McCall.

1000 Friends supports the city's UGB as well as strong land-use planning measures throughout the state through the language of "livability." Although the group places this on a continuum with sustainability (which includes compact neighborhoods and walkable communities), it sees much of the livability agenda as unrelated to sustainability (PDX01). In recent years, the original rural–urban coalition has become fractured, as urban development policies originally seen as necessary by both have become harder to sell to rural voters (PDX01, PDX04). As such, 1000 Friends has shifted its vision from the urban landscape to one that is more "people-oriented" (PDX01). 1000 Friends thus promotes land-use controls such as Portland's UGB by arguing that they directly contribute to quality of life. It sees livability as a process, exemplified by "the -er model": it is about making development smart*er* and neighborhoods dens*er* rather than using a set yardstick (PDX01).

5 The city's UGB is a constant source of debate, discussion, and interest in the city, and has led to artistic explorations, such as David Oates's ruminations about the boundary through walks along its edge (Oates, 2006).

1000 Friends does not use the language of sustainability, which it has found too vague for its advocacy purposes (PDX01). At the same time, its efforts to promote dense, livable urban form in Portland through comprehensive planning have been linked to urban sustainability (Godschalk and Anderson, 2012; Jepson and Edwards, 2010; Wheeler, 2004). As other actors in Portland suggested, "sustainability" is not always the best door to "walk through" in order to build support for sustainability policies (PDX04). 1000 Friends can thus be seen as an important advocate for related policies, even if it does not contribute to general awareness and interest in sustainability as such.

A second important group in Portland is the Coalition for a Livable Future (CLF), "a diverse partnership of organizations and individuals working together to create a more equitable and sustainable Portland metropolitan region" (CLF website, 2013). Similar to the *Zukunftsrat* in Hamburg, CLF is a network organization that seeks to coordinate the efforts of its member groups in order to increase their effectiveness in policy debates. CLF's main purpose is joining numerous smaller initiatives in order to build a larger, more coherent advocacy coalition for a common agenda. Thus, much of its work is to find common cause for interests of individual members. (1000 Friends is a member organization.) Unlike the *Zukunftsrat*, CLF's key mission is equity; it has had less success with the term "sustainability," especially in communities of color[6] (PDX02).

The group's work focuses on several issues that cut across policy sectors, like environmental and social justice, transportation, and equity. This allows the coalition to work with a diverse group of partner organizations like the Urban League and others. CLF prefers the "three 'E's" approach to sustainability, as it includes equity, an important part of their mission that is often lost in current discussions on sustainability (PDX02). The term "sustainability" has become less critical and useful to the group's mission in recent years; while I was conducting my field research in 2012, the group was discussing removing the term from its mission statement (PDX02).

Sustainability is not a term with coalition-building appeal for CLF. One interviewee worried that "sustainability is a really slippery term": people project their definition onto it, which is limiting, although it also presents a valuable opportunity to educate people (PDX02). It is also important to explain that sustainability is not just another term for environmentalism (PDX02). Part of CLF's mission is to better connect the environmental justice and environmental movements. Although "no organization has said we just want environmental quality in white, middle-class neighborhoods," this is what often happens based on how resources are distributed in some areas (PDX02).

---

6 My interview partner shared a concern from an activist in the African American community to make this point. This activist was cited as expressing dislike for the term sustainability because she did not consider the current situation in African American communities as something that should be "sustained."

Table 7.5 Outside sustainability groups in Portland

Organization	Purpose
1000 Friends of Oregon	• state-level advocacy organization founded by former Oregon governor to support growth management and comprehensive planning efforts • originally a coalition of rural and urban residents interested in protecting rich farmland of Willamette Valley; now focuses more on people to re-engage rural coalition members and gain new supporters
Coalition for a Livable Future	• coalition of social equity and sustainability groups in Portland with about 40 core members and about 50 supporting or affiliate groups • purpose is to join forces to develop larger coalition that supports all members' issues and works strategically toward goals • focuses on linking environmental and environmental justice movements and advocating together for key issues seen as important by all members

Portland does not have a sustainability-minded group like Hamburg or Seattle that uses the term "sustainability" to advance an explicit sustainability agenda. However, the work of 1000 Friends of Oregon and the Coalition for a Livable Future can be considered supportive of sustainability at the city and state level, even if they do not use this terminology. Because of their hesitance to use the word "sustainability," these groups do not enter definitional debates. Instead, they argue for the importance of related issues—land-use planning and equity—within the political arena. Both of these issues are important components of how sustainability has been operationalized in Portland's strategic plan, which suggests that these groups have had an impact. At the same time, their reluctance to use sustainability highlights its potential limitations in motivating local action in Portland.

Seattle: Mainstreaming Sustainability

Seattle has perhaps the most institutionalized and successful sustainability-minded groups among the six cities studied here. The city is home to two groups focused explicitly on sustainability issues in Seattle and the Pacific Northwest, Sustainable Seattle and the Sightline Institute. These organizations have done much to advocate for sustainability principles to be incorporated into policymaking. Sightline in particular has done work on several fronts to help frame local policy problems in the language of sustainability and create or use information and argumentative techniques to evaluate or advocate for policy through a sustainability lens. These groups have been supported by a rich network of foundations, such as the Bullitt Foundation and the Sustainable Path Foundation, that have supported sustainability initiatives in the region through grant programs explicitly focused on urban sustainability. The general impression seen in Seattle is one in which a rich ecosystem of sustainability groups work together to incorporate sustainability into policymaking and other decision-making processes (Pettibone, 2012).

Sustainable Seattle and the Sightline Institute have both made significant efforts in promoting sustainability in Seattle, both by defining the term itself and highlighting its importance in considering local policy issues. In this way, these organizations are quite similar to the *Zukunftsrat* in Hamburg. However, unlike the *Zukunftsrat*, both appear to have become more stable institutions in the city. According to its website, Sightline currently has 11 full- or part-time staff, compared to one part-time administrator at the *Zukunftsrat*. This allows them to devote resources to numerous research projects, events, and outreach on a larger scale than is possible for the *Zukunftsrat*.

The Sightline Institute was founded as Northwest Environment Watch in 1993 by Alan Durning, an inaugural member of Sustainable Seattle. The group calls itself a "think-and-do tank" focused on promoting sustainability in the Cascadia bioregion in the Pacific Northwest, which includes Seattle, Portland, and Vancouver, British Columbia in Canada. As discussed in Chapter 6, one of the group's early efforts was the Cascadia Scorecard, which measures sustainability trends in the region and includes distance-to-target figures. However, Sightline has de-emphasized such work over time as it has acknowledged the limited ability of such numbers to tell its story of sustainability. As one employee puts it:

> You can look at transit ridership and one set of people can say, 'These numbers show that transit is a terrible investment.' [While] others say, 'These numbers show that transit is a great investment.' They're both looking for the numbers that support their argument. But the problem is, if you're actually trying to be serious about it and not just win a political argument, but actually try to figure out what the data says, sometimes it's not clear. The point is not just to get more data, but to use what's already there and use it in the right way. Even when you have a lot of data, it doesn't necessarily always tell you a clear story.... So there's always a layer of storytelling or interpretation that has to happen, in order to make sense of the numbers you're seeing.
>
> Me: But one thing that I find interesting about Sightline's work is that it is taking data that is useful and using it to tell a story about sustainability.
>
> Interviewee: That's right, that's what we try to do. (SEA08)

Storytelling is the main focus of Sightline, which does so with reports, blog posts, and events. Through these channels, the organization seeks to provide information to sway local "influentials" and "mega-megaphones," those "people who are at the top of the information chain" (SEA08). Several staff members have ongoing projects, such as "Making Sustainability Legal," which focuses on little-known policies and regulations that outlaw simple lifestyle changes to sustainability, such as local ordinances against clotheslines (Durning and de Place, 2012). The impact of such work is difficult to measure, but Sightline's reports have been cited by other local media outlets, the organization enjoys good relations with local policymakers, and its small-scale actions have seen grassroots change.

In addition to Sightline Institute, Sustainable Seattle focuses on sustainability issues in Seattle. Originally formed by a group of dedicated volunteers, the organization has undergone several transformations in recent years (Holden, 2006). As discussed in the last chapter, one early project was to develop an indicator set that defined sustainability in a comprehensive, measurable way. The group's indicator report, published during development of Seattle's comprehensive plan in the 1990s, was supported by the city's planning department. This provided a foundation for the plan's use of sustainability as its central framework.

Since then, Sustainable Seattle has struggled to survive as an organization, changing its mission several times (Holden, 2006). According to Holden's (2006) ethnographic study of the group, after initial successes with its indicator set, which led to international diffusion aided by presentations at conferences around the world in the 1990s, several key original members left the group.[7] This led to funding troubles and a loss of mission. However, Sustainable Seattle has since developed new projects: the group recently spun off a happiness indicator project into an independent organization. Currently, Sustainable Seattle focuses on helping local businesses to be more sustainable, in partnership with the sustainable business program at the Bainbridge Graduate Institute (SEA06). In addition, the group's work has led directly or indirectly to the creation of numerous other groups working on sustainability-related issues in the region, such as the Sightline Institute, One Northwest, EcoBuilders, and the Environmental Coalition of South Seattle (Mitra, 2003, 23). The group's changing mission suggests successful completion of early projects and evolution into new areas of sustainability in Seattle.

Seattle appears to have benefited greatly from the existence of several foundations dedicated to funding sustainability projects. For example, the Bullitt Foundation is committed to and served as a key early donor to Sustainable Seattle (Holden, 2006). The Bullitt Foundation has "historically primarily funded mostly activist groups that were trying to educate the public and influence public policy," leading it to focus on issues like pollution clean-up and the salmon population

7 Although these departures were difficult for Sustainable Seattle, they may have benefited sustainability governance in Seattle. Richard Conlin became a city councilmember who advocated for incorporation of the group's indicators into city decision-making; Alan Durning founded what is now the Sightline Institute; Alan AtKisson is now a sustainability consultant in Sweden who has written several books on sustainability (e.g, AtKisson, 2009).

(SEA09). In recent years, however, the foundation has shifted to supporting urban sustainability projects in the region, particularly with a focus on defining urban sustainability:

> We're trying to find grantees that are in the university context or a think-tank context that can help us figure out what ... a sustainable city [would] really look like, so that when we take various steps, we're fairly confident that they are steps in the right direction: they won't come around and boomerang us. So we continue to fund ... still principally activist groups, but maybe a quarter, maybe a third of our grants now are going to people who are helping us sketch out the parameters of where we're going and how we should measure our progress. (SEA09)

The Bullitt Foundation thus sees a lack of clarity in terms of precisely what actions would make the city more sustainable and is taking steps to better understand what such actions would be. In addition, the foundation's new headquarters was built to meet the Living Building Challenge, an ambitious sustainable building code (International Living Future Institute, 2013).

In total, there are roughly 25–30 foundations active in the Seattle area that work together on environmental issues by looking for different leverage points at the same time (SEA10). One of these, the Sustainable Path Foundation, is a smaller organization that provides seven to 10 grants per year, including recent funding for Cascadia Green Building Council on green water infrastructure, especially stormwater, by using more plants and a decentralized approach; Climate Solutions to help municipal governments figure out how to transfer to a smart grid; and Earthcore to conduct urban forest monitoring with citizen scientists and create an online database (SEA10). The diverse funding streams available in Seattle and the number of different projects highlights the broad efforts of different groups involved in sustainability in the city.

Seattle presents a unique case among the cities studied in that it has several organizations dedicated to urban sustainability in the city and the region. In addition to the work of organizations such as the Sightline Institute, whose mission is similar to the *Zukunftsrat* in Hamburg, Seattle is home to several foundations that provide critical financial support for local sustainability projects and research. This has allowed sustainability-minded groups to flourish, in contrast to German cases where such organizations are dependent on grants from cash-strapped city governments. Indeed, this factor represents one of the greatest differences between case cities' experience with sustainability in the United States and Germany, and is discussed in the analysis.

Table 7.6 Outside sustainability groups in Seattle

Organization	Purpose
Sightline Institute	• think tank interested in advancing sustainability in the Cascadia bioregion
Sustainable Seattle	• nonprofit organization focused on building connections between local business actors on sustainability • grew out of volunteer group that developed world's first sustainability indicator set in 1990s and advocated for sustainability in Seattle
Bullitt Foundation	• foundation focused on funding urban sustainability projects in the Cascadia bioregion
Sustainable Path Foundation	• funds smaller sustainability projects in Puget Sound region

Analysis

As stated in the introduction, this chapter examines the efforts of outside sustainability groups to realize urban sustainability governance. The last two empirical chapters revealed the importance of one type of group in particular in development and use of strategic plans and sustainability indicators. This chapter looks broadly at the work of civil society groups engaged in urban sustainability, but focuses on the work of sustainability-minded groups in particular: *How do sustainability-minded groups use plans and indicators to promote urban sustainability governance?* This section begins by developing a typology of organizations working on sustainability-related issues at the city level based on the case study findings. It then discusses in more detail the work that sustainability-minded and other groups play in urban sustainability governance.

Developing a Typology of Organizations Working on Urban Sustainability

This chapter focuses on the work of outside sustainability groups in advocating for sustainability at the local level. However, as the six case cities demonstrate, local organizations can serve different purposes, and some hesitate to label their focus "sustainability." As such, it is helpful to develop a typology of these organizations and the work they do. I found three main categories of organizations working in some way on issues related to sustainability, based on the groups studied here:

- *sustainability-minded groups:* groups dedicated to defining sustainability in the local context and advancing sustainability in policy and public debate using the language and principles of sustainability, these are the groups that are the main focus of analysis; for example, Sightline Institute, *Zukunftsrat.*

- *coalition partners:* groups dedicated to change compatible with sustainability principles, but who do not themselves use the language of sustainability; for example, NYC EJA, Coalition for a Livable Future.
- *project-oriented groups:* groups committed to sustainability but less interested in engaging with political debates, who instead focus on concrete projects or technical research to advance sustainability; for example, IFEU, Sustainable Flatbush.

Although all three of these groups are important to realizing urban sustainability governance, the first appears to play a special role in increasing local political motivation for sustainability, particularly by reacting to strategic plans and using sustainability indicators. Sustainability-related advocates attempt to drive political motivation as well, but do not connect this activity to sustainability.

How Sustainability-minded Groups Contribute to Urban Sustainability Governance

A key shortcoming to successful use of strategic plans was a lack of motivation of key decision-makers. Sustainability-minded groups appear to contribute to urban sustainability governance by seeking to increase local motivation toward a sustainability transition, by advancing sustainability principles within the local context and challenging policies and visions they see as incompatible with those principles. In addition, although these groups' understanding of sustainability is broadly compatible with each other and the principles presented in Chapter 2, the different emphases may make it difficult to form a single coalition. However, the existence and work of these groups suggests that further study could reveal key insights into the nature of sustainability in the political world. At any rate, it is clear that these groups play an important role in the realization of urban sustainability governance.

One of the most important findings of this research is the role played by sustainability-minded groups in advancing sustainability language and concerns in local political debate. In this way, these institutions are instrumental in promoting the incorporation of sustainability principles into political decision-making. Such organizations, whether considered advocacy groups or think tanks, perform several types of important work. Specifically, they 1) bring the language and principles of sustainability to the local discourse; 2) link sustainability to local context; 3) build public support and engagement on sustainability issues; and 4) challenge incompatible political definitions and city visions from a sustainability perspective. These four functions are discussed in turn.

First, sustainability-minded groups bring the language and principles of sustainability to the public discourse. Sustainability-minded groups represent the vanguard of sustainability thinking in a city. They use indicators, publications, and public discussions to introduce the concept of sustainability to decision-makers and the public. In this way, they create awareness of the frame of sustainability and the role it can play in understanding and dealing with local policy issues.

Second, these groups link sustainability principles to the local context. This function is incredibly important and, like the others, insufficiently understood. While academic research and more abstract discussion on sustainability can help distill its core principles, these are often difficult for policymakers to grasp. How should a city planner respect "ecological limits" in designing a new residential development? Sustainability-minded groups translate sustainability principles into local issues in order to help policymakers understand how to incorporate these principles into decision-making processes.

Third, they build public support and engagement on sustainability issues. This can be through public events and initiatives to discuss sustainability and determine the community's key priorities, but it also includes sustainability indicators, publications, and local awards for sustainable individuals, businesses, and other initiatives. Sustainability-minded groups, but also related advocates, work to build the local discourse coalition related to sustainability.

Finally, sustainability-minded groups police the discourse. Numerous scholars have worried that sustainability, because it is a contested concept, can be defined at will by those who use it. This is true, but does not mean it should be abandoned. Indeed, all political concepts are contested and have been redefined over time, often into meanings far from their original intent.[8] In addition, if sustainability were to be abandoned because of its popularity or lack of specificity, whatever took its place would need to be defined and defended as well. From a discourse perspective, it is important to understand that discourse is simply part of policymaking. For this reason, sustainability-minded groups play a critical role in policing the discourse on sustainability by calling out definitions and policies they see as incompatible with sustainability principles. Their success is linked to their reputation as serious, thoughtful keepers of the sustainability flame; their actions important to counter-balance efforts to define sustainability in ways advantageous to the local government. Here, the *Zukunftsrat* presents the best example: its work challenging the city's growth vision led to the incorporation of sustainability indicators into a later draft and has fueled a debate on growth that in ongoing. Although not all groups studied here took up these functions, they were a core aspect of sustainability-minded groups' mission. In cities where such organizations did not exist (such as New York), political actors appeared to have far greater control over how sustainability is defined and realized politically in the local context. In cities with sustainability-minded groups, the term has been used as a site for debate over the city's fundamental normative vision (as in Hamburg) and a guiding principle for political decision-making (as in Seattle).

---

8 An excellent example of this is the term "meritocracy," which was originally coined as a warning against the sort of society in which members were given privileged status based on education and intelligence but has become an aspirational term for a society where wealth is linked to merit (Young, 2001).

Groups varied widely in terms of the strategies used to advocate for sustainability or the aspects of sustainability of interest to them. Whereas Heidelberg, Hamburg, and Seattle have used sustainability indicators as an important tool, network organizations in Portland and Hamburg seek to coordinate members' efforts. Sightline Institute has done perhaps the most original research, seeking to frame local policy issues through a sustainability lens, although groups in Hamburg have used sustainability to challenge local policy as well. Publications are a key tool used in these groups' advocacy, making them similar to think tanks in their pursuit of a particular political agenda. At the same time, the different traction groups have gotten in different cities suggests either the importance of a particular approach (such as type of publication or outreach), or their limited ability to affect political motivations in the short term. It is also possible that other factors, such as political culture, impact the influence of these groups.

The shape of the sustainability coalition also varies widely in the six cities studied. Although this is largely due to the different mix of organizations in each city to begin with (for example, the presence of environmental justice or international development groups), it also reflects the interest and ability of existing local groups to form partnerships. Although the Coalition for a Livable Future in Portland poses an excellent model of joining diverse organizations in order to increase their ability to advocate for a shared agenda, it is not clear what organizations could join a similar coalition in New York or Heidelberg, for example. In Heidelberg, the recently opened *Welthaus* that aims to promote joint projects between sustainability, environmental, and development groups in the city has not yet managed to develop a joint initiative in that city. This also means that not all of the cities studied were found to have sustainability-minded groups. Groups concerned with aspects of sustainability but not interested in calling themselves sustainability advocates *per se* were also less likely to engage in policing activities, although they sought to influence sub-discourses related to sustainability. Here, research on social capital and social movements could provide clues as to why different types of organizations exist in different cities.

Although sustainability has a fairly coherent definition among the groups studied, the focus can vary dramatically. In New York and Portland, environmental justice was a key theme; in other cities, this was barely mentioned. Groups in Portland, Seattle, and Hamburg were keenly interested in urban planning and land-use development. German organizations expressed more interest in international responsibility, particularly related to the global South. Although all of these issues can be connected to the definition of sustainability presented in Chapter 2, the different emphases in each city suggest both different local contexts and a specific way of dealing with trade-offs between sustainability principles in each city.

The fact that these groups did not exist in all the cities studied highlights their main shortcoming. Although sustainability is supposedly an overarching issue that should engage all residents and stakeholders in a city (and globally) in defining the future they want to see, it is not generally understood as such. Several cities I studied had only a few local actors engaged in sustainability. In addition, engagement came overwhelmingly from the environmental movement, particularly in Germany. It may be difficult if not impossible to develop a group dedicated to local sustainability in a city without passionate advocates. At the same time, the experience in cities with sustainability-related work, such as New York, suggests that it may be possible to empower groups working on sustainability issues even if the term itself holds little meaning. There is much work to be done in the areas of communication and coalition building in order to get those people and organizations whose work is related to sustainability but who do not explicitly identify with it working in concert with sustainability advocates. What these findings suggest is that urban sustainability governance may require some way of creating interest in forming such organizations in cities where they do not exist.

Funding: A Key Difference between the U.S. and Germany

The last two chapters did not find substantial differences between case cities based on country. However, the experiences of outside sustainability groups did differ between the United States and Germany, in terms of funding and the general atmosphere related to civil society groups. In general, nonprofit and civil society groups faced more fertile ground for funding in the United States, due to the number of foundations and private donations to such groups. In Seattle, for example, numerous foundations dedicated to funding local sustainability projects has helped organizations like Sustainable Seattle and the Sightline Institute grow and hire paid staff. By contrast, Berlin 21 and the *Zukunftsrat* in Hamburg have been almost completely reliant on city and project funding, putting both groups in a more precarious financial position and forcing them to rely almost exclusively on volunteers. These groups' experiences are reflective of general differences in how civil society is organized in these two countries (McGann and Johnson, 2005; Stone, 1996). But Heidelberg has found one possible solution: paying local groups to offer free environmental consulting in order to provide an important administrative service for the city and guarantee organizations a stable funding source. If these groups are to exist and thrive in the long-term, the key issue of funding should be examined more closely.

Conclusions

This chapter highlighted the important role played by sustainability-minded groups in shaping local discourse and action related to sustainability. The groups presented in this chapter showed the wide spectrum of functions: providing information related to sustainability in the city context, advocating for a sustainability framework to understand local policy issues, and challenging city visions incompatible with their understanding of sustainability. Outside groups can work with government by providing information, analysis, and policy, or by undertaking local projects, or challenge government policies or its very definition of sustainability. Sustainability can be used as an umbrella term to strengthen collaboration and highlight cross-cutting aspects of important policy issues. But it can also be an awkward term that does not mobilize groups.

Because of the important work outside groups perform in incorporating sustainability principles into policymaking, a comprehensive framework for urban sustainability governance needs to include sustainability-minded groups. These organizations have not yet been the subject of adequate scientific study. Future research could focus on the following issues: 1) the different roles played by these organizations, how these roles are determined, how they change; 2) the impact of their work on policymaking; 3) their interactions with other actors; and 4) funding and long-term sustainability of the institutions themselves.

al stakeholders. Sustainability can be used as an umbrella term to strengthen collaboration and highlight cross-cutting aspects of many policy issues. But it can also be a devalued term that does not mobilize anyone.

Because of the importance of [illegible] groups in [illegible] sustainability [illegible] policymakers [illegible] sustainability [illegible] [illegible] [illegible] [illegible] [illegible] [illegible] and how [illegible] sustainability [illegible]

PART III
Realizing Sustainability Principles in Urban Sustainability Governance

Chapter 8
Reconceptualizing Urban Sustainability Governance

Part II presented case study analysis on the role played by three factors in incorporating sustainability principles in six U.S. and German cities. Both strategic plans and sustainability indicators can contribute to urban sustainability governance, but do not necessarily perform their expected roles. Instead, problems can arise in two key areas. First, use of these instruments was strongly influenced by actors' motivation: high-level interest in sustainability is thus a prerequisite for plans and indicators to be effective tools. Second, the instruments themselves may not be defined in a way consistent with sustainability. More often, the plans studied here sought to promote growth rather than address environmental and social concerns. Indicators can help operationalize sustainability and link strategic goals to key performance measures, but must be developed in conjunction with motivated users to do so. Sustainability-minded groups appear to address this issue, by defining sustainability in the local context, pushing for action, and critiquing incompatible activities. However, the fact that these groups only existed in a few of the cities studied here demonstrates a barrier to their value in other cities.

This chapter synthesizes the main findings from the six case studies and highlights the implications for urban sustainability governance, from both a theoretical and practical perspective. The first section distills the main findings from the three empirical chapters to present urban sustainability governance as requiring both governance instruments and outside groups for distinct but interconnected purposes. It also highlights the importance of actors' motivations for using sustainability in the local context. In the second section, I discuss how the sustainability principles seen in the academic literature manifested in the case studies as well as what new issues emerged. Finally, I discuss the implications of these findings for the study and practice of sustainability governance. Specifically, I argue that: 1) sustainability principles should not be considered a key motivation of political actors, even in cities labeled as sustainability leaders, without additional evidence; 2) more research needs to be conducted on the motivational aspect of urban sustainability governance, such as the role played by sustainability-minded groups; and 3) advocates interested in effecting a sustainability transition need to devote more attention to building interest in sustainability more generally.

The Promise and Limitations of Plans, Indicators, and Groups

The empirical analysis in Part II presented in detail the various roles that strategic plans, sustainability indicators, and sustainability-minded groups play in incorporating sustainability principles into political decision-making processes in six U.S. and German cities. Table 8.1 presents an overview of the main findings from these chapters.

Table 8.1 The role of plans, indicators, and sustainability-minded groups in urban sustainability governance in six U.S. and German cities

City	Strategic plans	Sustainability indicators	Outside sustainability groups
Berlin	• Plan developed by deliberative external LA21 process, saw no further use • New plan developed by urban development minister dedicated to growth	• Indicators developed by external LA21 initiative, saw no further use	• Coordinate LA21 outreach • Network grassroots groups, projects • Hold public events
Hamburg	• Plans developed by political coalitions called for growth, later tempered to include environmental and social concerns	• Indicators used for advocacy, to challenge political strategy	• Challenge political strategy/ narrative, present sustainability vision • Network local groups • Maintain indicators
Heidelberg	• Plan developed through deliberative city-led process, still in use	• Indicators linked to plan used to report • Sustainability test used to check sustainability of all major political decisions • Externally developed indicators not used	• Advocate for sustainability and development primarily in the Global South • Research institutes develop indicators for government agencies

New York	• Sustainability plan emerged from economic development goals: coordinates built environment and climate change strategy	• Indicators in plan used as performance measures	• Strong local environmental justice groups put EJ issues on political agenda • Grassroots sustainability groups focus on neighborhood projects
Portland	• State-mandated plan developed through deliberative city-led process over several administrations, implementation plan in development	• State-level indicators developed and maintained by external advisory group and not used; funding cut by political rivals • City-level indicators developed by external advisory group that has been disbanded	• Political coalition coordinates advocacy for livability issues • Advocate for continued growth management, smart growth policies
Seattle	• State-mandated plan uses sustainability framework, lack of follow-through on neighborhood engagement	• Multiple externally created indicator sets used for advocacy, one still in use but has become more ad-hoc	• Link local political issues to sustainability • Build coalition for sustainability effort • Maintain indicators

Source: Author's synthesis of empirical findings.

These six cities' experiences with plans, indicators, and outside groups differed widely, but some patterns emerge. First, the differences between the United States and Germany were generally negligible compared to the variations between the cities within these two countries, except in the area of civil society engagement. City size appeared a more important difference in terms of political motivation and engagement of outside groups. In addition, cities were seen to use sustainability in various ways—or not use it at all—in decision-making. Supporting Berke and Conroy's (2000) early findings that the use of sustainability terminology did not necessarily make strategic plans more "sustainable," this research shows that use of the term is not so easily connected to incorporation of sustainability principles into decision-making. For this reason, it is important to look beyond the terminology used in evaluating cities' efforts toward urban sustainability governance.

Both strategic plans and sustainability indicators have the capacity to promote sustainability. But in cases where motivation was lacking, their promise was not realized. In particular, strategic plans not linked to statutory requirements—which usually included conditions that made the plans more conducive to sustainability governance—did not necessarily use sustainability as a guiding framework or define sustainability in a way linked to its key principles. More often, the strategic plans studied here were used to address growth, whether by actively promoting it (Berlin, Hamburg, New York) or accommodating it (Heidelberg, Portland, Seattle). In addition, passage of a plan or development of indicators is insufficient to realize urban sustainability governance if actors are not motivated to actually use these instruments. Building and maintaining motivation of local decision-makers to use sustainability is at least as important as developing appropriate governance instruments.

Sustainability indicators also have a capacity to incorporate sustainability principles into decision-making. Their use within government appears to consist primarily of operationalizing a strategic vision and measuring performance on related initiatives. Here, the indicators included in New York's *PlaNYC* have helped city agencies and their partners work together on the city's numerous sustainability initiatives. In addition, Heidelberg's sustainability test has forced decision-makers to consider the sustainability ramifications of key political decisions. Outside groups were key indicator users; their work focused on building motivation seen as being a key factor for the use of strategic plans. Hamburg's HEINZ indicator set has been a mainstay of the *Zukunftsrat*'s advocacy work, serving to present its definition of sustainability and advocate for policies to address low-performing indicators. Sightline Institute and Sustainable Seattle used indicators for similar reasons.

Therefore, the role played by sustainability-minded groups in seeking to motivate these actors, by connecting sustainability to the local political context and challenging visions incompatible with its principles, appears to be an important complement to governance instruments. Evidence of this role was seen in cities with such groups, particularly Seattle and Hamburg. In addition to trying to increase the salience of sustainability in the local context, sustainability-minded groups were also active in challenging political definitions of sustainability or city visions that they saw as incompatible with sustainability. This was seen most clearly in the numerous critical publications released by local civil society groups in Hamburg. The way the *Zukunftsrat* challenged Hamburg's growth-centric vision led to inclusion of sustainability goals in a subsequent draft. Such work may be helpful in preserving the term's meaning.

At the same time, the fact that only two of the cities studied (Hamburg and Seattle) have active sustainability-minded groups implies that this is only one of numerous tasks performed by civil society actors interested in urban sustainability governance. Berlin, for example, has strong grassroots engagement on sustainability issues, but Berlin 21 has shifted its efforts in recent years away from the political sphere toward helping to network and build the grassroots efforts taking place at the neighborhood level. The Coalition for a Livable Future in Portland does political advocacy work on livability and environmental justice issues related to sustainability, but has at times eschewed the term "sustainability" itself as unhelpful to its work. Both advocacy and project work appear important, so it may be interesting to study further how each can be nurtured and how organizations such as Sightline affect the composition of local political coalitions and the project work of urban gardeners and lending libraries.

The three factors studied are likely to be most effective when they work together to contribute to urban sustainability governance. Both strategic plans and sustainability indicators can help incorporate sustainability principles into political decision-making. But these instruments are less effective in cases where local actors are not motivated to work toward a sustainability transition. Sustainability-minded groups can increase motivation by telling stories that link sustainability to the local context. Each can serve a particular function, but none can achieve sustainability governance alone.

A critical factor seen throughout was the role of motivation. This suggests the importance of additional research into values change. Actors whose worldviews were consistent with sustainability principles were more likely to use the instruments studied here to incorporate those principles into decision-making. However, in many cases those in positions that could be instrumental in incorporating sustainability into policymaking had little understanding of or interest in the concept. As the case of Heidelberg showed, sustainability cannot simply be administrated: it must be linked to normative values of local decision-makers and communities. For this reason, the engagement of sustainability advocates whose values are compatible with sustainability principles and who are able to translate those principles into action in the local policy context is essential.

Much of the action taken here appeared to be led by dedicated advocates, whether in the form of active mayors (Heidelberg, New York), advocates in sustainability-minded groups (Berlin, Hamburg, Seattle), or administrators (Hamburg, Portland). Therefore, it may be necessary to acquire or train new actors to play these important roles if values or motivation are found to be intransigent. Such change can be spurred by funding for outside groups, interest at the state level, or a dedicated champion in a position of power. Urban sustainability governance requires more of these factors to become present in more cities. The variety of levers suggests multiple points of action to spur sustainability transition.

It appears from the findings here that sustainability advocates can indeed translate sustainability principles into the local policy context and argue for their importance from a moral or practical standpoint. Indeed, the lack of such advocates in some of the case cities was found to be an important factor inhibiting incorporation of sustainability principles into local decision-making processes. For example in New York, some grassroots groups and individuals working on sustainability issues were unhappy with the strategy put forward by *PlaNYC*, but did not make a concerted effort to challenge or change it. In New York especially, power appears to play a role in privileging certain definitions over others, simply by providing powerful actors (such as Mayor Bloomberg) more resources to state their case. Here, actor groups that formed coalitions to present a consistent critical message appeared the most successful in challenging the visions of dominant players.

Why did cities use strategic plans and sustainability indicators? Generally, these instruments were used more to aid decision-making and promote a political vision than to transition to urban sustainability governance specifically. For this reason, these instruments require motivation on sustainability as a prerequisite for their effectiveness in incorporating sustainability principles into political decision-making. As such, sustainability-minded groups—who link sustainability to the local context and advocate for its incorporation into decision-making—may be a missing piece by explicitly focusing on this area. Table 8.2 summarizes how each factor analyzed can contribute to urban sustainability governance as well as its limitations.

Table 8.2 The role of three key factors in urban sustainability governance

	Strategic plans	Sustainability indicators	Sustainability-minded groups
Purpose, benefits (+)	• Can help link sustainability to city's strategic political vision	• Can operationalize sustainability principles within local context • Can link sustainability goals to policymaking • Have diverse uses to match local needs	• Can connect sustainability principles to local context • Can make sure that local action is consistent with sustainability principles
Requirements, limitations (–)	• Requires institutional support (for example, legislative requirements) to succeed • Does not necessarily match sustainability principles	• Only relevant if indicators match users' values	• Require engaged advocates, stable financial resources • Do not exist in all cities

This book used comparative case study analysis to better understand urban sustainability governance in high-consumption countries. However, despite the different national-level contexts in the United States and Germany presented in Chapter 3, major differences in use of instruments were not seen between the two countries. Cities in both countries sought to address sustainability issues through strategic plans and use of sustainability indicators, but the most successful plan implementation was seen in the U.S. cities of Portland and Seattle, where state-level planning requirements and support aided city efforts. The German strategic plans studied here were generally not linked to federally required land-use and zoning plans, suggesting that they could learn from the Pacific Northwest in this area.

In addition, cities in both countries saw similar use of sustainability indicators. Whereas Heidelberg and New York developed performance measures to measure implementation of their strategic plans, sustainability-minded groups in Seattle and Hamburg developed indicator sets to advocate for inclusion of sustainability in local decision-making. In general, indicators in the United States were slightly more likely to be developed by external advocacy groups while indicators in Germany were developed in conjunction with government projects, but evidence of both were seen in each country.

Perhaps the most important difference between the two countries was the interaction of actors on sustainability issues. Although both Hamburg and Seattle have sustainability-minded groups with a similar purpose, the way they operate has diverged dramatically in recent years. Local groups in the United States focus on building awareness and action within civil society, and can draw heavily on sources of funding from like-minded foundations and private donations. In Germany, outside groups are more likely to be funded through government channels. When this funding source dries up, as with Berlin 21, it can jeopardize the group's future. For this reason, German sustainability groups rely more heavily on volunteers than those in the United States. At the same time, national-level advisory boards in Germany enjoy close proximity to decision-makers. Each approach has strengths and weaknesses; further research could explore these in more detail and perhaps suggest lessons for groups in each country.

Sustainability Principles and Urban Sustainability Governance

In order to understand the contributions of different factors to urban sustainability governance, it is important to evaluate them against some definition of sustainability. This study defined sustainability as an essentially contested concept with five core principles: acknowledgment of limits, long time scales, use of complexity, deliberation, and reflexivity (see Chapter 2). Urban sustainability governance is thus the extent to which these principles are incorporated into political decision-making processes. This section examines how well these principles were reflected at the city level and issues of operationalization that arose (see Table 8.3).

Table 8.3 Incorporating sustainability principles into political decision-making

Principle	Supporting evidence	Contradictory evidence	Further work
Acknowledgment of limits	• Greenhouse gas inventories, other indicators • Urban growth boundaries, growth management efforts • Supportive language in vision documents (for example, reference to international responsibility)	• Pro-growth policies (primarily population) • Efficiency efforts not tied to net reductions in throughput	• More academic work is needed to link global ecological limits to city-level processes and institutions • Particularly, work needed to connect population flows to limits discourse
Long time scales	• Comprehensive planning efforts, planning mandates • Incorporation of long-term effects into policymaking	• Lack of plan institutionalization • Prioritization of policies with short-term effects that contradict cities' long-term strategy	• Many cities could benefit from state or federal level mandates for long-term planning
Systems thinking	• Strategic placement of sustainability office • Interconnection of indicators, policy sectors • Cross-agency collaborations and public–private partnerships	• Individual agencies' or external actors' ability to circumvent strategic plan	• Further institutionalization of sustainability/strategic planning office • Better understanding of various effects of policies
Deliberative	• Participatory planning processes • Advisory and neighborhood councils • Active engagement of sustainability advocates	• Participatory element at times tacked on to predesigned plan • Inclusion does not necessary indicate empowerment	• More academic work on the role of different types of participation in governance • Experimentation with different deliberative methods • Institutional support
Reflexive	• Plan updates and changes reflective of new circumstances • Regular collection of sustainability indicators with scorecards and dashboards	• Lack of plan updates • Lack of reflection on or acknowledgment of previous plans	• Sunset provisions for policies • Institutionalized plan updates or evaluative processes • Better understanding of the role of plans and indicators in governance

These findings show that evaluating urban sustainability governance through sustainability principles is useful. The analytic approach highlights problems with research that simply describes cities' development of sustainability plans or indicators, as the terminology used does not necessarily reflect the instrument's or actors' goals. For this reason, it is important to develop evaluation criteria for urban sustainability governance. These five principles were an initial attempt that could be further developed.

In general, cities did best at incorporating long-term thinking into policy processes. This is no surprise, as urban planning is perhaps the most forward-thinking policy tool in common use (for example, Miller, 2004). However, their success with this principle depends to some extent on how it is understood: is a 20-year strategic plan truly long-term or merely medium-term planning? Considering the difficulties of projections, this principle may need to be reformulated. One possibility could be the consideration of future effects of policies, particularly in the face of changing circumstances, such as demographic change.

The cities studied here also generally tried to incorporate elements of systems thinking, reflexivity, and deliberation into decision-making processes, although the results indicate that there is still important work to be done in these areas. For example, although five of the six case cities (Hamburg being the exception) used some form of public or stakeholder engagement in developing a strategic plan, the heterogeneity of approaches suggests that the purpose and methods of incorporating deliberation are still unclear. In addition, deliberation entails open-ended outcomes, which suggests that whomever is to be included be consulted early and be able to affect the content of the city vision. In some cases, engagement consisted more of participation or information-sharing than true deliberation. At the same time, the cities that sought to include the public (Heidelberg and Portland) required more time and resources to guide the planning process than the stakeholder approach used in New York. It could thus be useful to develop a typology of different deliberative tools and their effect on policy processes. As participatory methods are more resource-intensive and time-consuming than top-down decision-making, it is important for deliberative processes to add value to decisions. In addition, deliberative decision-making raises questions about the legitimacy of decisions—whom to include and how such that decisions are democratic and reflect local values—that require urgent consideration.

Complexity, like long time scales, was difficult to measure conclusively. Although all cities used plans and indicators to integrate policy sectors and interconnections were mentioned by many, it is unclear whether this is a full incorporation of this principle. Complexity is best seen in systems approaches such as urban metabolism; as such, it may be better understood as a tool to understand sustainability, but difficult to evaluate in the policymaking context. The truly important measure—the placement of sustainability within the city decision-making apparatus—could likely be defined better in other terms, such as Grabow et al.'s (2011) understanding of sustainability as a top priority.

Reflexivity was seen especially in plan and indicator updates that showed trends and revised goals or initiatives. Some cities, such as New York, Portland, and Seattle, demonstrated ongoing commitments to their plans. At the same time, reflection sometimes meant changing course, as seen by Sustainable Seattle. After its indicators were considered for the city's strategic plan, the organization shifted to other priorities. Legislative requirements can aid reflexivity, such as regular reporting or state-mandated planning. In some cases, however, such as Berlin's current strategic plan development, actors have failed to consider previous work, making it difficult to learn from past mistakes.

The principle least incorporated into decision-making in the six case cities was the backbone of sustainability: an acknowledgment of limits. This is due to some extent to the difficulty of defining limits in an open system (Rees, 1997), as well as to the challenge of linking urban processes to ecological systems. Research into urban ecology and metabolism is beginning to address this (for example, Moore, Kissinger, and Rees; 2013; Pincetl, 2012; Zhang, 2013); in addition, efforts such as the ecological footprint can help frame individual consumption in a way that motivates policy. However, several case cities pursued a goal that may be incompatible with the principle of limits: population and economic growth. The *Zukunftsrat* in Hamburg has done the most work to challenge the notion that population growth through pursuit of highly qualified migrant labor is sustainable, but negative reactions in New York and Heidelberg, as well as the established growth management strategies in Portland and Seattle, have also highlighted this question. It is possible that the policy of poaching migrants from other cities is indeed unsustainable, as it improves one city's economy at the expense of another's (Granholm and Mulhern, 2011). At the same time, New York linked its growth strategy to sustainability by arguing that New York City residents had dramatically lower carbon emissions than those in surrounding suburbs (City of New York, 2007; see also Owen, 2009). Indeed, when considered from an ideological standpoint, urban growth strategies are closely linked to neoliberal ideology (Hackworth, 2007). Sustainability, which challenges many other aspects of neoliberal doctrine, may thus find urban growth another point of contention. This issue is worth a closer look.

In general, the principles used here served as a useful tool to analyze the content and process of strategic plans and sustainability indicators. However, the analysis that stems from using abstract principles to evaluate urban sustainability governance is subjective to some extent. This may be impossible to avoid when faced with a normative term such as sustainability. At the same time, it shows how undefined key elements within sustainability remain, such as deliberation or complexity. Related academic literatures could be tapped to better conceptualize these principles. Equally important, however, is additional work to define sustainability governance in the urban context. Here, academic institutions could work together with sustainability-minded groups to pool different streams of information relevant to the particular city of interest.

Implications for Sustainability Science and Practice

One underlying assumption of much of the sustainability governance literature is that cities (that is, political actors in cities) *want* to incorporate sustainability principles into decision-making processes. Thus, a main focus of research focuses on developing policy instruments for sustainability governance—such as strategic plans and sustainability indicators—with the implication these are the key ingredient for a sustainability transition. In addition, many often take sustainability initiatives at face value: if a city calls a particular project a "sustainability plan," its implementation, or even simply its passage, is taken as evidence that the city is becoming more sustainable.

This research has found two main problems with these assumptions. First, it found that the creation of a plan or indicator set said little about its use or influence in urban decision-making. Berlin's current urban development minister appears to be unaware of the city's Local Agenda 21; indicator sets in Berlin, Heidelberg, and Seattle have come and gone with little concrete impact on policymaking. Second, due to the contested nature of sustainability, plans labeled "sustainability plans" (as in New York) may be less interested in sustainability than comprehensive plans that generally eschewed sustainability terminology (as in Portland) (which supports Berke and Conroy, 2000).

Sustainability is far less settled than is generally acknowledged, even by authors who call it a contested concept (for example, Connelly, 2007). At the same time that academic debate examines the implications of ecological limits for economic growth, sustainability is being used to promote patterns of urban development that may be incompatible with academic definitions. The political meaning of sustainability in the cities studied was often divorced from the academic literature. This was due to at least three reasons: 1) inadequate linkage between an academic understanding of sustainability and the urban context; 2) insufficient understanding of sustainability among political actors; and 3) different values, worldviews, and motivation. Although further research can deal with the first issue, the other two require active engagement at the local level, a domain that may be more suitable for local sustainability-minded groups.

The first problem found in cities was that the general definition of sustainability often does not translate directly into guidance for a distinct policy program in cities. For example, what does sustainability say about transportation priorities: should a neighborhood develop high-speed rail or a new bus line? Indeed, sustainability seems not to provide the answers to complex political questions, but rather to offer questions to guide deliberation. In the neighborhood's case, it would offer decision-makers and residents criteria to help inform their decision. This requires, as Fürst (2002) and Grabow et al. (2011) have argued, efforts to put sustainability into the heads of decision-makers. Although several—primarily German—authors have called for this, no guidance exists in the sustainability literature on how this is to be achieved. One promising possibility found here is provided by sustainability-minded groups, which sought to operationalize sustainability in the local political context. Some evidence of political action was seen by their actions. Further research could analyze factors that affect these groups' political impact.

Sustainability-minded groups may also be able to address the second problem: lack of clarity among political actors of what sustainability is. At a sustainability event in Heidelberg, the city's mayor and state science minister revealed this when answering a question about how they have incorporated sustainability into their daily lives (Würzner at "*Lebendiger Neckar*" event, Jun. 15, 2013). The mayor provided a stock environmentalist answer: he bikes to work and has solar panels on his roof. The minister's response was even more telling: she said she "doesn't have time for sustainability," as if it were an extra burden, but that she buys organic food. Even at a meeting of sustainability advocates in Berlin in late 2013, the moderator asked whether sustainability was only for rich people—essentially framing sustainability as an elite consumer choice, despite the fact that unsustainability is essentially a problem caused by the rich world—the premise of the question was not challenged by any of the six panelists (Berlin 21 "*Soziale Stadt*" event, Nov. 28, 2013). Sustainability is a relatively young concept and by no means settled. Much work remains to begin to reach agreement on certain principles and approaches and develop a shared storyline (cf. Hajer, 1995).

Finally, this research found that a major impediment to incorporation of sustainability principles into political decision-making was that the values, worldviews, and motivation of decision-makers are different in many ways from those of sustainability advocates. A major concern of several administrations (for example, in New York, Hamburg, and Berlin) was the city's financial situation. Mayors in all three cities have developed aggressive policies to encourage highly skilled workers and investment from outside the city, which could be described as neoliberal (Hackworth, 2007). Sustainability could challenge this thinking in several ways, as several efforts by advocates in Hamburg demonstrate, but has not done so in a concerted way. Indeed, the sustainability literature has provided little guidance to date on how cities could become financially stable aside from asserting that balanced budgets represent "sustainable finance" (Grabow et al., 2011). For sustainability to be politically relevant, it should provide solutions to urgent political problems.

This suggests that more conceptual work is needed to link urban concerns to sustainability principles. It also indicates the importance of education and advocacy. Here, advocates played an important role in translating the concept into the local context. Groups studied in this research that perform such work—primarily in Seattle and Hamburg—affected local policy discussions in important ways. In addition, academics could provide more concrete examples of how principles can be realized in actual policy contexts. Finally, it appears important to find better analytic frameworks to link use of the term "sustainability" to adherence to sustainability principles. To that end, the discussion here of what those principles are may improve the analytical sophistication of future research on urban sustainability governance.

Chapter 9
Conclusions

This study explored the role that strategic plans, sustainability indicators, and sustainability-minded groups play in urban sustainability governance, which is conceptualized as incorporating sustainability principles into political decision-making processes. I presented sustainability as an essentially contested concept guided by five core principles: acknowledgment of natural limits, long time scales, systems perspective, deliberation, and reflexivity. I used this conceptualization in an empirical study of six U.S. and German cities to understand the extent to which two popular governance instruments—strategic plans and sustainability indicators—aid in this process.

First, Chapter 5 showed that cities do not necessarily use strategic plans to promote sustainability; more often, this instrument is used to deal with local growth issues. Plans can be used to promote sustainability principles, but only if actors are motivated by sustainability. Chapter 6 highlighted the various ways that sustainability indicators can be used toward urban sustainability governance. These uses vary by actor group; in general, indicator sets are only effective if developed in conjunction with target users. I distilled three key uses of indicators in political decision-making: advocacy, strategic, and administrative. Advocacy use and the role of civil society groups have not been adequately incorporated into conceptual frameworks of indicator use.

These outside groups' role in urban sustainability governance thus became the focus of Chapter 7. That chapter explored the role of outside sustainability groups in urban sustainability governance, finding three key types of organizations. Sustainability-minded groups focus on defining sustainability in the local context and advancing the concept in policy and public debates. Similarly, coalition partners are dedicated to change compatible with sustainability principles, but do not use the language of sustainability. Finally, project-oriented groups are committed to sustainability at the grassroots or academic level and are less interested in political discussions. Particularly the first group may play an important in increasing motivation for sustainability sometimes seen as lacking in development of strategic plans and indicators.

This chapter briefly discusses the implications of these findings and further areas for research.

Implications: Motivating Actors for Urban Sustainability Governance

Several key findings from this study have implications for both researchers and practitioners. First, strategic plans and sustainability indicators are not sufficient to achieve a sustainability transition at the urban level when motivation is lacking. Thus, further research on how to increase political motivation and interest in sustainability is essential to ensure a transition to urban sustainability governance in more cities.

Much of the research effort in terms of urban sustainability governance is focused on instruments and institutions, which ignores motivational factors demonstrated here as a key prerequisite for those instruments' success. Although sustainability can be made "*Chefsache*" (a top priority) by developing such governance instruments, key actors must be committed to their use for them to have value. In addition, more work is needed to understand how to better link local definitions of sustainability to broader principles; otherwise, the term will slowly become meaningless. Sustainability advocates, particularly in the form of sustainability-minded groups, appear to be an important actor in this regard, as they both link local problems to a story of sustainability and contest local definitions and actions at odds with sustainability principles.

Another important finding was the tendency of political actors in the cities studied to define sustainability in line with growth principles. This could be due to several factors: path dependency that leads decision-makers to stick with business-as-usual; normative frameworks arisen from education and training during a period in which neoliberal ideas became popular; a tendency of decision-makers to support powerful interests that are often aligned with growth policies. Whatever the reason, the goals of especially the big cities studied (that is, Berlin, Hamburg, and New York) tended to be more interested in pursuing growth than sustainability; in some cases (particularly New York), sustainability was used to justify growth policies. However, in cases where the public was more involved in planning deliberations (such as Portland and Heidelberg), this orientation was questioned and at times changed. This suggests an ideological rift between the public and political elites on growth questions that indicates the influence more open and deliberative planning and governance processes could have on city visions and policy. Additional research could further explore this dynamic.

Outside groups play an important role in defining sustainability in the local context. However, not all groups appear to understand how to contribute to urban sustainability governance in a meaningful way. The example of externally developed urban sustainability indicator sets highlights this point. Organizations that developed external indicator sets in the hopes that they would be directly taken up by decision-makers were disappointed in every case analyzed here (e.g, in Berlin and Heidelberg). By contrast, groups that developed indicators in order to advocate for sustainability were able to influence the policy process, albeit in ways that are difficult to quantify (especially Hamburg and Seattle). Further work is needed on how sustainability-minded groups use indicators, the extent to which their use of indicators is successful in motivating political actors, and what factors affect their influence.

What is the role of academics? The understanding of urban sustainability governance that emerges from this analysis is one in which civil society actors are actively engaged in defining sustainability and linking it to local problems in their city. These findings support Lafferty's (2004) depiction of sustainability as an "outside-in" political vision: as such, more attention is needed to understand how outside actors seek to convince local decision-makers to take up the cause of sustainability.

This suggests that academics interested in making a difference in the transition to sustainability should do more than simply conduct and publish research. Academic research is not used one-to-one in political decision-making, but, as numerous scholars have pointed out over the past several decades, policymakers pick and choose the information that best fits their worldview and political goals. For this reason, it is not enough to simply create more data; what is needed is new storylines to demonstrate how scientific findings require policy change. In other words, scientific research must be translated into the local political context, made meaningful to decision-makers and the public. It may no longer be enough for researchers to conduct niche research and publish in academic journals. Further work could explore how researchers can connect to local sustainability groups both to determine a politically relevant research agenda and to discuss important findings.

Urban sustainability governance thus requires the constant attention of outside advocates to motivate decision-makers and challenge incompatible action. This has several important implications:

- Defining sustainability requires constant effort, through information, education, and discussion. Sustainability-minded groups perform important work here but are far from universally present in cities.
- More work is required to link sustainability principles to the local political context. The research on sustainability tends to focus on diffuse, global concerns, such as ecological limits, that are difficult to link to urban problems. Again, sustainability-minded groups have done excellent work in arguing for certain local policies using the language and principles of sustainability. Researchers could aid this effort by thoughtfully examining issues such as urban growth, demographic change, and the built environment from a sustainability perspective.
- Although sustainability must be defined through deliberation, the definition presented bases it on five core principles. Even for a contested concept, it is important that some principles remain generally fixed. This requires engagement: definitions that fail to meet certain requirements must be challenged. More work could be done to better understand how to do so effectively.

- Finally, some governance instruments can stimulate ongoing discussion on sustainability issues. Legally mandated comprehensive plans that require regular updates, as well as institutionalized sustainability tests, indicators, and reporting can aid this.

Realizing Urban Sustainability Governance: Next steps

This research has highlighted motivation as an important aspect of urban sustainability governance. Specifically, it is important for key decision-makers to agree with sustainability principles in order to incorporate them into political decision-making processes. This conceptualization suggests important practical and academic work.

Practical Work

All actors at the city level should be engaged in sustainability transition. However, this research has shown the special role played by sustainability-minded groups in promoting sustainability principles and linking them to the local political context. The efficacy of these groups could be improved by several actions.

First, the few sustainability-minded groups that do exist should serve as leading examples of important ideational work in relation to urban sustainability governance. Leading groups such as Sightline Institute—perhaps in collaboration with other sustainability advocates—could distill key lessons and tools into publications to aid development of similar efforts in new cities. Local sustainability advocates would do well to dedicate at least some effort to the long-term mission of defining sustainability. Here, think tanks and research institutes could work together to develop resources to translate sustainability into the political and urban context and clearly outline tangible goals. More research is needed to better understand what aspects of their work are most successful and how to increase the impact of these groups, especially in cases of limited funding. Funding institutions interested in advancing sustainability governance should direct resources to these groups.

Second, these groups need to be better connected to the broader sustainability movement. In Germany, the Council for Sustainable Development or other national-level organization could facilitate networking between city groups, as well as between the city and state level. Sustainability-minded groups appear to play an important role in advocating for urban sustainability governance and would benefit from stronger connections to local universities conducting research on relevant issues. Civil society groups with similar missions may benefit from coalition-building efforts, such as the work conducted by the Coalition for a Livable Future in Portland. Here, local context is important: the idea of sustainability is in some cases more meaningful when divorced from the term itself.

This last issue is critical for sustainability advocates. Because sustainability is a contested concept, its meaning is derived from use over time. Although this study does not examine the long-term effects of using a particular terminology, it does find that powerful actors in cities define the term to meet their needs. It is thus important for advocates to look beyond the headlines and consider the values being advanced. The five principles outlined in Chapter 2 provide a first step to such critical examination. Defining sustainability, however, should be considered an ongoing endeavor.

Areas for Further Research

In addition to practical issues raised, this research points to several areas for future study. There are several important questions to address in the coming years, which I present here as an idealized research agenda:

- How robust are the five principles of sustainability presented here, in particular outside the U.S. or European context? Can they be translated, for example, to the global South?
- How can key features unique to cities be analyzed through a sustainability lens? In particular, how do population and urban economic growth fit into sustainability? Do different types of migration or growth policies have different implications for sustainability? How can the principles of sustainability be better operationalized in the urban context?
- Can sustainability be understood as a political ideology? What can the history of other political ideologies, such as neoliberalism, teach us about the ideational component of a sustainability transition? What insights can be gained from discourse theory and theories of paradigm shift or social movements that could be applied to a sustainability transition?
- Does this initial understanding of sustainability-related groups hold up to further research? What role do different types of organizations play in urban sustainability governance, as well as beyond the city level and outside the United States and Germany? How do different groups contribute to incorporation of sustainability principles in political decision-making processes?

These questions are by no means exhaustive but can, I hope, illuminate some possible next steps on the path to urban sustainability governance.

Bibliography

Abgeordnetenhaus Berlin. (2006). *Lokale Agenda 21: Berlin zukunftsfähig gestalten.* (2006). Berlin. Retrieved December 7, 2011, from http://www.stadtentwicklung.berlin.de/agenda21/de/service/download/agenda_21_web_2.pdf.

Abbott, C. (2001). *Greater Portland: Urban Life and Landscape in the Pacific Northwest.* Philadelphia, PA: University of Pennsylvania Press.

Abbott, C. (2004). Centers and edges: Reshaping downtown Portland. In: Ozawa, C.P., Ed. *The Portland Edge: Challenges and Successes in Growing Communities.* Washington, DC: Island Press, 164–83.

Abbott, C. (2011). *Portland in Three Centuries: The Place and the People.* Corvallis, OR: Oregon State University Press.

Acuto, M. (2013). City leadership in global governance. *Global Governance,* 19(3), 481–98.

Adger, W.N. and Jordan, A., eds. (2009). *Governing Sustainability.* Cambridge: Cambridge University Press.

Agyeman, J. and Evans, T. (2003). Toward just sustainability in urban communities: Building equity rights with sustainable solutions. *The Annals of the American Academy of Political and Social Science,* 590(1), 35–53. doi:10.1177/0002716203256565.

Altrock, U. (2004). Das Leitbild von der 'Wachsenden Stadt' – genial oder fatal? In: Altrock, U. and Schubert, D., eds. *Wachsende Stadt: Leitbild – Utopie – Vision?* Wiesbaden: VS Verlag, 77–94.

Altrock, U., Knaap, G., and Cohen, J. (2004). Die Smart-Growth-Bewegung in den USA – Vorbild für intelligentes Wachstum in Deutschland? In: Altrock, U. and Schubert, D., eds. *Wachsende Stadt: Leitbild – Utopie – Vision?* Wiesbaden: VS Verlag, 117–26.

Altrock, U. and Schubert, D., eds. (2004). *Wachsende Stadt: Leitbild – Utopie – Vision?* Wiesbaden: VS Verlag.

Amt für Stadtentwicklung und Statistik. (2009). *Auswertung der Nachhaltigkeitsprüfung von Gemeinderatsvorlagen in den Jahren 2007 und 2008.* Anlage 1 zur Drucksache 0199/2009/IV, 2009. Heidelberg: Stadt Heidelberg. Available at: http://ww1.heidelberg.de/buergerinfo.

Amt für Statistik Berlin-Brandenburg. (2012). *Kernindikatoren zur nachhaltigen Entwicklung Berlins: Datenbericht 2012.* Potsdam: Amt für Statistik Berlin-Brandenburg. Retrieved February 25, 2014, from https://www.statistik-berlin-brandenburg.de/home/pdf/Nachhaltigkeit_barrierefrei_31_10.pdf.

Amt für Statistik Berlin-Brandenburg. (2014). *Die kleine Berlin-Statistik 2013.* English edition. Potsdam: Amt für Statistik Berlin-Brandenburg. Retrieved December 15, 2014, from https://www.statistik-berlin-brandenburg.de/produkte/kleinestatistik/AP_kleinestatistik_en_2013_be.pdf.

Angotti, T. (2008). Is New York's sustainability plan sustainable? Paper presented to the *joint conference of the Association of Collegiate Schools of Planning and Association of European Schools of Planning (ACSP/AESOP)*, Chicago, July 2008. Sustainability Watch working paper. Retrieved January 16, 2012, from http://www.hunter.cuny.edu/ccpd/repository/files/is-nycs-sustainability-plan-sustainable.pdf.

Angotti, T. and Checker, M. (2011, May). Keeping watch on the plan for a greener New York. *Gotham Gazette.* Retrieved March 2, 2012, from http://www.gothamgazette.com/article/Environment/20110511/7/3524.

AtKisson, A. (2009). *The Sustainability Transformation: How to Accelerate Positive Change in Challenging Times.* London: Earthscan.

Baatz, C. (2013). Responsibility for the past? Some thoughts on compensating those vulnerable to climate change in developing countries. *Ethics, Policy & Environment*, 16(1), 94–110.

Bagley, K. and Gallucci, M. (2013). *Bloomberg's Hidden Legacy: Climate Change and the Future of New York City.* Inside Climate News e-book. Kindle edition.

Ballon, H., Ed. (2012). *The Greatest Grid: The Master Plan of Manhattan.* New York: Museum of the City of New York.

Barg, S., Anielski, M., and Waddell, J.T. (2006). Using performance information in government budgeting and reporting: Review of best practices. Retrieved April 1, 2014, from http://www.iisd.org/pdf/2006/measure_performance_info.pdf.

Bartlick, S. (2014, 27 May). Berlin voters claim Tempelhof. *Deutsche Welle.* Retrieved January 15, 2015, from http://www.dw.de/berlin-voters-claim-tempelhof/a-17663944.

Bauriedl, S., Schindler, D., and Winkler, M., (Eds). (2008). *Stadtzukünfte denken: Nachhaltigkeit in europäischen Stadtregionen.* Munich: oekom Verlag.

Beatley, T. (1995). Planning and sustainability: The elements of a new (improved?) paradigm. *Journal of Planning Literature*, 9(4), 383–95. doi:10.1177/088541229500900405.

Beauregard, R.A. (2006). *When America Became Suburban.* Minneapolis, MN: University of Minnesota Press.

Beck, G. (2012). *Agenda 21.* New York: Threshold Editions.

Becker, J. (2005). Measuring progress towards sustainable development: An ecological framework for selecting indicators. *Local Environment*, 10(1), 87–101.

Behörde für Stadtentwicklung und Umwelt [BSU]. (2001). *Kursbuch Umwelt: Ziele für ein zukunftsfähiges Hamburg.* Freie und Hansestadt Hamburg. Retrieved May 31, 2013, from http://www.isebek-initiative.de/uploads/sn/FHH_2001_UB_KursbuchUmweltHamburg.pdf.

Behörde für Stadtentwicklung und Umwelt [BSU]. (2012). *Umweltprogramm 2012 bis 2015: Ziele und Maßnahmen.* Freie und Hansestadt Hamburg. Retrieved March 13, 2015, from http://www.hamburg.de/contentblob/3529414/data/umweltprogramm-2012-2015.pdf.

Bell, S. and Morse, S. (2001). Breaking through the glass ceiling: Who really cares about sustainability indicators? *Local Environment*, 6(3), 291–309. doi:10.1080/1354983012007328.

Bell, S. and Morse, S. (2008). *Sustainability Indicators: Measuring the Immeasurable?* Second edition. London: Earthscan.

Bell, S., Eason, K., and Frederiksen, P. (eds). (2011). *A Synthesis of the Findings of the POINT Project*. Retrieved July 12, 2011, from http://www.point-eufp7.info/storage/POINT_synthesis_deliverable 15.pdf.

Berg, B.F. (2007). *New York City Politics: Governing Gotham.* New Brunswick, NJ: Rutgers University Press. Kindle edition.

Berger, K. (2009). *Pugetopolis: A Mossback Takes on Growth Addicts, Weather Wimps, and the Myth of Seattle Nice.* Seattle: Sasquatch Books.

Berke, P.R. and Conroy, M.M. (2000). Are we planning for sustainable development? An evaluation of 30 comprehensive plans. *Journal of the American Planning Association*, 66(1), 21–33.

Berlin 21, Ed. (2014). *Vielfältig nachhaltig für Berlin: Auf dem Weg in eine zukunftsfähige Hauptstadt.* Munich: oekom.

Berliner Zeitung. (2013, 31 May). Berlin muss Millionen zurückzahlen. *Berliner Zeitung* online. Retrieved April 14, 2014, from http://www.berliner-zeitung.de/berlin/volkszaehlung-zensus-2011-berlin-muss-millionen-zurueckzahlen,10809148,23080874.html.

Berlin Senat. (2009). *Bericht zur Lokale Agenda 21 Berlin.* Drucksachen 15/3245 and 15/5221. Retrieved April 29, 2010, from http://www.stadtentwicklung.berlin.de/agenda21/de/service/download/Bericht_LA21Berlin2009.pdf.

Bernt, M. (2012). The "double movements" of neighbourhood change: Gentrification and public policy in Harlem and Prenzlauer Berg. *Urban Studies*, 49(14), 3045–62. doi:10.1177/0042098012437746.

Beuermann, C. (1998). Germany: Five years after Rio and still uphill all the way? In: Lafferty, W.M. and Eckerberg, K., eds. *From the Earth Summit to Local Agenda 21: Working Towards Sustainable Development.* International Environmental Governance series: Vol. 12. London: Earthscan, 106–39.

Beuermann, C. (2001). Germany: Regulation and the precautionary principle. In: Lafferty, W.M. and Meadowcroft, J., eds. (2001). *Implementing Sustainable Development: Strategies and Initiatives in High Consumption Societies.* Oxford: Oxford University Press, 85–111.

Biermann, F. (2013). Curtain down and nothing settled: Global sustainability governance after the "Rio+20" Earth Summit. Environment and Planning C: Government and Policy, 31(6), 1099–114. doi:10.1068/c12298j.

Blanco, H. (2012). Public participation in neighborhood planning, a neglected aspect of community livability: The case of Seattle. In: Wagner, F. and Caves,

R., eds. *Community Livability: Issues and Approaches to Sustaining the Well-Being of People and Communities.* London: Routledge, 183–97.

Block, T., Van Assche, J., and Goeminne, G. (2013). Unravelling urban sustainability: How the Flemish city monitor acknowledges complexities. *Ecological Informatics*, 17, 104–10.

Bomberg, E. (2009). "Governance for sustainable development: The United States and the European Union compared." In Schreurs, M.A., Selin, H., and VanDeveer, S. (eds), *Transatlantic Environment and Energy Politics: Comparative and International Perspectives*. Farnham: Ashgate, 21−40.

Bonneville Power Administration [BPA]. (n.d.). *About us.* Website Retrieved May 14, 2014, from http://www.bpa.gov/news/AboutUs/History/Pages/default.aspx.

Bossel, H. (1999). *Indicators for Sustainable Development: Theory, Method, Applications*. Winnipeg, Canada: International Institute for Sustainable Development. Retrieved May 10, 2010, from http://www.iisd.org/pdf/balatonreport.pdf.

Bossel, H. (2001). Assessing viability and sustainability: A systems-based approach for deriving comprehensive indicator sets. *Ecology and Society*, 5(2), 12. Retrieved May 10, 2010, from http://www.consecol.org/vol5/iss2/art12.

Bowers, C.A. (2000). *Let them Eat Data: How Computers Affect Education, Cultural Diversity, and the Prospects of Ecological Sustainability.* Athens, GA: University of Georgia Press.

Brambilla, R. and Longo, G. (1980). *What Makes a City Livable? Learning from Seattle.* Institute for Environmental Action.

Brand, K-W. and Jochum, G. (2000). *Der deutsche Diskurs zu nachhaltiger Entwicklung* (No. MPS-Texte 1/2000). Retrieved February 13, 2013, from www.sozialforschung.org/wordpress/wp-content/uploads/2009/09/kw_brand_deutscher_nachh_diskurs.pdf.

Brand, K-W. and Fürst, V. (2002). Sondierungsstudie: Voraussetzungen und Probleme einer Politik der Nachhaltigkeit – Eine Exploration des Forschungsfelds. In: Brand, K-W., ed. *Politik der Nachhaltigkeit: Voraussetzungen, Probleme, Chancen – eine kritische Diskussion.* Berlin: Edition Sigma, 15–109.

Bressers, H.T.A. (2004). Implementing sustainable development: How to know what woks, where, when and how. In: Lafferty, W.M., ed. *Governance for Sustainable Development: The Challenge of Adapting Form to Function.* Cheltenham: Edward Elgar, 284–318.

Brugmann, J. (1997). Is there a method in our measurement? The use of indicators in local sustainable development planning. *Local Environment*, 2(1), 59–72. doi:10.1080/13549839708725512.

Brulle, R.J. (2000). *Agency, Democracy, and Nature: The U.S. Environmental Movement from a Critical Theory Perspective.* Cambridge, MA: MIT Press.

Bryner, G.C. (2001). The United States: 'Sorry—not our problem.' In: Lafferty, W.M. and Meadowcroft, J., eds. *Implementing Sustainable Development:*

Strategies and Initiatives in High Consumption Societies. Oxford: Oxford University Press, 273–302.

Bulkeley, H., Jordan, A., Perkins, R., and Selin, H. (2013). Governing sustainability: Rio+20 and the road beyond. *Environment and Planning C: Government and Policy*, 31(6), 958–70. doi:10.1068/c3106ed.

BUND Hamburg, Diakonie Hamburg und Zukunftsrat Hamburg, eds. (2010). *Zukunftsfähiges Hamburg: Zeit zum Handeln*. A study by the Wuppertal Institut für Klima, Umwelt, Energie. Munich: Dölling und Galitz Verlag.

BUND and Misereor, eds. (1996). *Zukunftsfähiges Deutschland: Ein Beitrag zu einer global nachhaltigen Entwicklung. Studie des Wuppertal Instituts für Klima Umwelt Energie GmbH*. Basel: Birkhäuser Verlag.

Bundesministerium für Verkehr, Bau und Stadtentwicklung [BMVBS]. (2007). *Towards a National Urban Development Policy in Germany: Memorandum*. Berlin. Retrieved April 4, 2014, from http://www.nationale-stadtentwicklungspolitik.de/nn_251568/Content/Publikationen/NSP/memorandum_towards_a_national_urban_policy,templateId=raw,property=publicationFile.pdf/memorandum_towards_a_national_urban_policy.pdf.

Bundesregierung. (2002). *Perspektiven für Deutschland: Unsere Strategie für eine nachhaltige Entwicklung*. Berlin, Germany. Retrieved May 21, 2010, from http://www.bundesregierung.de/nsc_true/Content/DE/_Anlagen/2006-2007/perspektiven-fuer-deutschland-langfassung,templateId=raw,property=publicationFile.pdf/perspektiven-fuer-deutschland-langfassung.

Bundesregierung. (2004). *Perspektiven für Deutschland: Unsere Strategie für eine nachhaltige Entwicklung. Fortschrittsbericht 2004*. Berlin, Germany. Retrieved May 21, 2010, from http://www.bundesregierung.de/Webs/Breg/nachhaltigkeit/Content/_Anlagen/fortschrittsbericht-2004.html.

Bundesregierung. (2008). *Perspektiven für Deutschland: Unsere Strategie für eine nachhaltige Entwicklung. Fortschrittsbericht 2008*. Berlin, Germany. Retrieved May 21, 2010, from http://www.bundesregierung.de/Content/DE/Publikation/Bestellservice/_Anlagen/2008-11-17-fortschrittsbericht-2008.html.

Bundesstiftung Baukultur. (2014). Über Baukultur. Bundesstiftung Baukultur website. Retrieved April 16, 2014, from http://www.bundesstiftung-baukultur.de/informationen/ueber-baukultur.html.

Bureau of Planning and Sustainability [BPS]. (n.d.). *Comprehensive Plan Update Working Draft*. BPS website. Retrieved July 24, 2013, from https://www.portlandoregon.gov/bps/60986.

Bureau of Planning and Sustainability [BPS]. (2013a). *City of Portland Comprehensive Plan Working Draft: Part 1*. City of Portland. Retrieved July 24, 2013 from https://www.portlandoregon.gov/bps/article/429864.

Bureau of Planning and Sustainability [BPS]. (2013b, 8 Apr). *My Portland Plan: How is the Portland Plan being Implemented through the Comprehensive Plan?* BPS website. Retrieved July 24, 2013, from http://www.portlandonline.com/portlandplan/index.cfm?a=443033&c=50730.

Bürgerschaft der Freien und Hansestadt Hamburg [BFHH]. (2010). *Leitbild Hamburg: Wachsen mit Weitsicht.* City of Hamburg. Drucksache 19/5474. Retrieved February 14, 2013, from http://www.hamburg.de/contentblob/2115436/data/download-objekt-leitbild.pdf.

Byles, J. and Kazi, O., eds. (2008). *The New York 2030 Notebook.* Notebook #01. New York: Institute for Urban Design.

Cass, D. (1965). Optimum growth in an aggregative model of capital accumulation. *Review of Economic Studies*, 32(2), 233–40.

Cass, L.R. (2007). Measuring the domestic salience of international environmental norms: Climate change norms in American, German and British climate policy debates. In: Pettenger, M.E., ed. *The Social Construction of Climate Change: Power, Knowledge, Norms, Discourses.* Farnham: Ashgate, 23–50.

City of Hamburg. (2012). *Hamburg Climate Action Plan 2007–2012: Update 2011–2012.* Retrieved April 15, 2014, from http://www.hamburg.de/contentblob/3959638/data/download-update-2011-12-englisch.pdf.

City of New York. (2007). *PlaNYC: A Greener, Greater New York.* New York. Retrieved January 12, 2012, from http://nytelecom.vo.llnwd.net/o15/agencies/planyc2030/pdf/full_report_2007.pdf.

City of New York. (2009). *PlaNYC Progress Report: A Greener, Greater New York.* New York. Retrieved January 15, 2012, from http://nytelecom.vo.llnwd.net/o15/agencies/planyc2030/pdf/planyc_progress_report_2009.pdf.

City of New York. (2011). *PlaNYC Update: A Greener, Greater New York.* New York. Retrieved May 15, 2012, from http://nytelecom.vo.llnwd.net/o15/agencies/planyc2030/pdf/planyc_2011_planyc_full_report.pdf.

City of Portland. (2012). *The Portland Plan: Prosperous. Educated. Healthy. Equitable.* Portland, OR. Retrieved May 30, 2012, from http://www.portlandonline.com/portlandplan/index.cfm?c=58776&a=398384.

Cobb, C., Halstead, T., and Rowe, J. (1995). If the GDP is up, why is America down? *The Atlantic* October 1995. Retrieved April 1, 2014, from http://www.theatlantic.com/past/politics/ecbig/gdp.htm.

Cobb, C.W. and Rixford, C. (1998). *Lessons to be Learned from the History of Social Indicators.* Redefining Progress. San Francisco, CA. Retrieved May 13, 2010, from http://www.rprogress.org/publications/1998/SocIndHist.pdf. doi:10.2307/492843.

Coenen, R. and Grunwald, A., eds. (2003). *Nachhaltigkeitsprobleme in Deutschland: Analyse und Lösungsstrategien.* Berlin: Edition Sigma.

Cohen, S. (2011). *Sustainability Management: Lessons from and for New York City, America, and the Planet.* New York: Columbia University Press.

Connelly, S. (2007). Mapping sustainable development as a contested concept. *Local Environment*, 12(3), 259–78.

Conrad, J. (2000). *Nachhaltige Entwicklung: einige begriffliche Präzisierungen oder der heroische Versuch, einen Pudding an die Wand zu nageln.* FFU Report 00-07. Forschungsstelle für Umweltpolitik, Freie Universität Berlin. Berlin. Retrieved

February 26, 2013, from http://www.polsoz.fu-berlin.de/polwiss/forschung/systeme/ffu/publikationen/2000/conrad_jobst_20004/rep_00-07.PDF.

Conroy, M.M. and Berke, P.R. (2004). What makes a good sustainable development plan? An analysis of factors that influence principles of sustainable development. *Environment and Planning A*, 36(8), 1381–96. doi:10.1068/a367.

Cowett, P. (2008). From potholes to planning: Will the city's transportation agency changing routes? Sustainability Watch working paper. Retrieved January 16, 2012, from http://www.hunter.cuny.edu/ccpd/repository/files/from-potholes-to-planning.pdf.

Creswell, J.W. (1998). *Qualitative Inquiry and Research Design: Choosing among Five Traditions.* Thousand Oaks, CA: SAGE Publications.

Daly, H.E. (1977). *Steady-state Economics.* San Francisco, CA: W.H. Freeman.

Daly, H.E. (1987). The economic growth debate: What some economists have learned but many have not. *Journal of Environmental Economics and Management*, 14(4), 323–36. doi:10.1016/0095-0696(87)90025-8.

Daly, H.E. and Cobb, J.B., Jr. (1989). *For the Common Good: Redirecting the Economy Toward Community, the Environment, and a Sustainable Future.* Boston: Beacon Press.

Davidson, K.M. (2010). Reporting systems for sustainability: What are they measuring? *Social Indicators Research*, 100(2), 351–65. doi:10.1007/s11205-010-9634-3.

Dearborn, N.W. (1983). Global 2000: Radar for the ship of state. *Futures*, 15(2), 111–25. doi:10.1016/0016-3287(83)90105-2.

De Haan, G. (2006). The BLK "21" programme in Germany: A "Gestaltungskompetenz"-based model for Education for Sustainable Development. *Environmental Education Research*, 12(1), 19–32. doi:10.1080/13504620500526362.

Department of Land Conservation and Development [DLCD]. (2007). *The Complete Planner's Guide to Periodic Review.* State of Oregon. Retrieved July 24, 2013, from http://www.oregon.gov/LCD/docs/publications/periodic_review_guide_100107.pdf.

Department of Planning and Development [DPD]. (2005). *Toward a Sustainable Seattle: City of Seattle Comprehensive Plan.* City of Seattle. Retrieved April 5, 2012, from http://www.seattle.gov/DPD/cs/groups/pan/@pan/documents/web_informational/dpdd016610.pdf.

Department of Planning and Development [DPD]. (2014). *Seattle 2035: Updating Seattle's Comprehensive Plan. City of Seattle.* Retrieved May 1, 2014, from http://www.seattle.gov/DPD/cs/groups/pan/@pan/documents/web_informational/p2112855.pdf.

Department of Sanitation. (2006). *Comprehensive solid Waste Management Plan.* New York City. Retrieved July 18, 2012, from http://www.nyc.gov/html/dsny/html/swmp/swmp-4oct.shtml.

De Place, E. and Kaleda, C. (2007). *Two Years of Measure 37: Oregon's Property Wrongs*. Seattle, WA: Sightline Institute. Retrieved May 30, 2012, from http://www.sightline.org/wp-content/plugins/download-monitor/download.php?id=39.

Dernbach, J.C. (2002). Learning from the President's Council on Sustainable Development: The need for a real national strategy. *Environmental Law Reporter*, 32, 10648–66.

Deutsche Welle. (2015, February 16). Hamburg SPD mayor Scholz wins big but now needs a partner. *Deutsche Welle*. Retrieved March 6, 2015, from http://www.dw.de/hamburg-spd-mayor-scholz-wins-big-but-now-needs-a-partner/a-18259959.

Diefenbacher, H., Karchner, H., Stahmer, C., and Teichert, V. (1997). *Nachhaltige Wirtschaftsentwicklung im regionalen Bereich. Ein system von ökologischen, ökonomischen und sozialen Indikatoren.* Forschungsstätte der evangelischen Studiengemeinschaft (FEST). Reihe A, Nr. 42. Heidelberg.

Diefenbacher, H., Teichert, V., and Wilhelmy, S. (2009). *Leitfaden: Indikatoren im Rahmen einer Lokalen Agenda 21 (Vierte Auflage).* Heidelberg: Forschungsstätte der Evangelischen Studiengemeinschaft e.V. Retrieved May 18, 2011, from http://www.nachhaltigkeitsbeirat-thueringen.de/files/aktuelles/leitfaden_Indikatoren LA21-4Aktualisierung2009.pdf.

Diefenbacher, H. and Zieschank, R. (2011). *Woran sich Wohlstand wirklich messen lässt: Alternativen zum Bruttoinlandsprodukt.* Munich: Oekom Verlag.

Dietz, S. and Neumayer, E. (2009). Economics and the governance of sustainable development. In: Adger, W.N. and Jordan, A., eds. *Governing sustainability.* Cambridge: Cambridge University Press, 259–82.

Dobson, A. (2009). Citizens, citizenship and governance for sustainability. In: Adger, W.N. and Jordan, A., eds. *Governing Sustainability.* Cambridge: Cambridge University Press, 125–41.

Dobson, A. (2014). Sustainability citizenship. *Open Citizenship*, 5(1), 86–95.

Dolowitz, D.P. and Medearis, D. (2009). Considerations of the obstacles and opportunities to formalizing cross-national policy transfer to the United States: A case study of the transfer of urban environmental and planning policies from Germany. *Environment and Planning C: Government and Policy*, 27(4), 684–97. doi:10.1068/c0865j.

Donahue, P. (2014, 11 Dec). Berlin mayor Mueller sworn in, ending Wowereit's 13-year run. *BloombergBusiness.* Retrieved March 7, 2015, from http://www.bloomberg.com/news/articles/2014-12-11/berlin-mayor-mueller-sworn-in-ending-wowereit-s-13-year-run.

Dovern, J., Rickels, W., and Quaas, M.F. (2012). *Nachhaltigkeitspotenzial deutscher Städte.* Kiel Policy Brief 50. Institut für Weltwirtschaft Kiel. Retrieved January 23, 2013, from http://www.ifw-kiel.de/wirtschaftspolitik/politikberatung/kiel-policy-brief/KPB_50.pdf.

Dresner, S. (2008). *The Principles of Sustainability.* Second edition. London: Earthscan.

Deutsche Bundesstiftung Umwelt [DBU]. (2007, 26 Sep). Umweltschutz, Wirtschaft und weniger Bürokratie können Hand in Hand gehen. Deutscher

Umweltpreis 2007: Einzelwürdigung Beate Weber. Retrieved April 1, 2014, from http://www.dbu.de/123artikel27151_555.html.

Du Pisani, J. a. (2006). Sustainable development – historical roots of the concept. *Environmental Sciences*, 3(2), 83–96. doi:10.1080/15693430600688831.

Durning, A. and de Place, E. (2012). *Making Sustainability Legal: Outdated Rules that Stop Affordable, Green Solutions.* Seattle, WA: Sightline Institute. Retrieved May 15, 2014, from http://www.sightline.org/download/2223.

Earth Engineering Center and Urban Habitat Project. (2001). *Life after Fresh Kills: Moving beyond New York City's Current Waste Management Plan.* Columbia University. Retrieved February 24, 2012, from http://www.seas.columbia.edu/earth/EEC-SIPA-report-NYC-Dec11.pdf.

Ehrenfeld, J.R. (2008). *Sustainability by Design: A Subversive Strategy for Transforming our Consumer Culture.* New Haven, CT: Yale University Press. Kindle edition.

empirica, Gesellschaft für Struktur- und Stadtforschung mbH & Stadt Heidelberg. (1995a). *Stärken-Schwächen-Analyse.* Heidelberg: City of Heidelberg. [bound volume also includes position statements on scenarios by various groups].

empirica, Gesellschaft für Struktur- und Stadtforschung mbH and Stadt Heidelberg. (1995b). *Szenarien zur Stadtentwicklung Heidelberg 2010.* Heidelberg: City of Heidelberg. [bound volume also includes position statements on scenarios by various groups].

Espinosa, A. and Walker, J. (2011). *A Complexity Approach to Sustainability: Theory and Application.* London: Imperial College Press.

Evans, B., Joas, M., Sundback, S., and Theobald, K. (2006). Governing local sustainability. *Journal of Environmental Planning and Management*, 49(6), 849–67. doi:10.1080/09640560600946875.

Farley, H.M. and Smith, Z.A. (2013). Defining sustainability: Refocusing a distorted concept. Paper presented at *ASPA Annual Conference, Governance & Sustainability: Local Concerns, Global Challenges.* New Orleans, 15–19 March.

Feitelson, E. (2004). The potential and limitations of planning in advancing sustainability notions at the sub-national level: An introduction. In: Feitelson, E., ed. *Advancing Sustainability at the Sub-national Level: The Potential and Limitations of Planning.* Aldershot: Ashgate, 1–9.

Feldman, E.J. (1981). *A Practical Guide to the Conduct of Field Research in the Social Sciences.* Boulder, CO: Westview Press.

Feller-Länzlinger, R., Haefeli, U., Rieder, S., Biebricher, M., and Weber, K. (2010). *Messen, werten, steuern: Indikatoren—Entstehung und Nutzung in der Politik. Eine Analyse mittels Fallbeispielen aus den Bereichen Nachhaltige Entwicklung und Bildung.* TA-SWISS-Studie TA-54/2010. Zentrum für Technologiefolgen-Abschätzung. Bern. Retrieved August 17, 2011, from http://www.ta-swiss.ch.

Ferro, S. (2014, 7 Apr). When branding is too good: A cautionary tale from New York's Citi Bike. *Fast Company: Design.* Retrieved April 9, 2014, from http://

www.fastcodesign.com/3028632/terminal-velocity/when-branding-is-too-good-a-cautionary-tale-from-new-yorks-citi-bike.

Fischer, F. (2000). *Citizens, Experts and the Environment* Durham, NC: Duke University Press.

Fischer, F. (2009). *Democracy and Expertise: Reorienting Policy Inquiry.* Oxford: Oxford University Press.

Fischer, F. and Black, M., eds. (1995). *Greening Environmental Policy: The Politics of a Sustainable Future.* New York: St. Martin's Press.

Fischer, F. and Hajer, M.A., eds. (1999). *Living with Nature: Environmental Politics as Cultural Discourse.* Oxford: Oxford University Press.

Flegenheimer, M. (2014, 26 Mar). Citi Bike system successful, but wobbly from the start. *The New York Times.* Retrieved April 9, 2014, from http://www.nytimes.com/2014/03/27/nyregion/as-citi-bike-approaches-anniversary-worries-abound.html.

Florida, R. (2004). *The Rise of the Creative Class: And How it's transforming Work, Leisure, Community, and Everyday Life.* New York: Basic Books.

Forest Park Conservancy. (2014). Organization's website. Retrieved February 1, 2014, from http://www.forestparkconservancy.org/forest-park.

Frank, S. (2008). Stadtentwicklung durch die EU: Europäische Stadtpolitik und URBAN-Ansatz im Spannungsfeld von Lissabon-Strategie und Leipzig Charta. *Raumforschung und Raumordnung*, 66(2), 107–17.

Freie und Hansestadt Hamburg [FHH]. (2002). *Leitbild: Metropole Hamburg – Wachsende Stadt.* City of Hamburg. Retrieved February 11, 2013, from www.wachsender-widerstand.de/wachsende_stadt.pdf.

Fürst, D. (2002). Schwierigkeiten der fachübergreifenden Koordination. In: Brand, K-W., ed. *Politik der Nachhaltigkeit: Voraussetzungen, Probleme, Chancen – eine kritische Diskussion.* Berlin: Edition Sigma, 179–91.

Gahin, R., Veleva, V., and Hart, M. (2003). Do indicators help create sustainable communities? *Local Environment*, 8(6), 661–6.

Galbraith, J.K. (1958). *The Affluent Society.* New York: Houghton Mifflin.

Gallie, W.B. (1956). Essentially contested concepts. *Proceedings of the Aristotelian Society*, 56, 167–98.

Gallucci, M. (2012, 12 May). Bill to ban sustainability and climate change action fails in Arizona. *Inside climate news* website. Retrieved May 15, 2014, from http://insideclimatenews.org/news/20120511/bill-ban-united-nations-agenda-21-sustainability-climate-change-global-warming-iclei-john-birch-society-kansas.

Garcia-Sanchez, I.M. and Prado-Lorenzo, J-M. (2008). Determinant factors in the degree of implementation of Local Agenda 21 in the European Union. *Sustainable Development*, 16, 17–34. doi:10.1002/sd.

Geffers, G. (2000). Steter Tropfen höhlt den Stein. In: GERMANWATCH (Eds), *Agenda 21: Hamburgs mühsamer Weg ins 21. Jahrhundert.* Dritter Nord-Süd WatchBericht. Hamburg: Konkret Literatur Verlag, 62–72.

Gehrlein, U., ed. (2000). *Wege zur Zukunftsbeständigkeit: Strategien und Instrumente zur Umsetzung des Leitbildes nachhaltiger Entwicklung.* Darmstädter interdisziplinäre Beiträge 3. Münster: agenda Verlag.

Gehrlein, U. (2004). *Nachhaltigkeitsindikatoren zur Steuerung kommunaler Entwicklung.* Wiesbaden: Verlag zur Sozialwissenschaften.

GERMANWATCH Regionalgruppe Hamburg (Eds). (2000). *Agenda 21: Hamburgs mühsamer Weg ins 21. Jahrhundert.* Dritter Nord-Süd WatchBericht. Hamburg: Konkret Literatur Verlag.

Gibbs, D. and Krueger, D. (2007). Containing the contradictions of rapid development?: New economy spaces and sustainable urban development. In: Krueger, R. and Gibbs, D., eds. *The Sustainable Development Paradox: Urban Political Economy in the United States and Europe.* New York: The Guilford Press, 95–122.

Glesne, C. (1999). *Becoming Qualitative Researchers: An Introduction.* Second edition. New York: Longman.

Godschalk, D.R. (2004). Land use planning challenges: Coping with conflicts in visions of sustainable development and livable communities. *Journal of the American Planning Association*, 70(1), 5–13.

Godschalk, D.R. and Anderson, W.R. (2012, Jan). *Sustaining Places: The Role of the Comprehensive Plan.* American Planning Association Planning Advisory Service Report No. 567.

Goldy. (2013, 11 Dec). Fuck the state: Seattle's secret plan to save metro bus service. *The Stranger.* Retrieved April 4, 2014, from http://www.thestranger.com/seattle/fuck-the-state/Content?oid=18454197.

Gordon, D.J. (2013). Between local innovation and global impact: Cities, networks, and the governance of climate change. *Canadian Foreign Policy Journal*, 19(3), 288–307. doi:10.1080/11926422.2013.844186.

Grabow, B., Bießwenger, K-D., Bock, S., Melcher, L., and Schneider, S. (2011). *Städte für ein nachhaltiges Deutschland: Gemeinsam mit Bund und Ländern für eine zukunftsfähige Entwicklung.* Part of the Lord Mayors "Sustainable City" dialogue. Deutsches Institut für Urbanistik im Auftrag vom Rat für Nachhaltige Entwicklung. Retrieved January 10, 2012, from http://www.nachhaltigkeitsrat.de/?id=6384.

Granholm, J. and Mulhern, D. (2011). *A Governor's Story: The Fight For Jobs and America's Economic Future.* New York: PublicAffairs.

Grober, U. (2010). *Die Entdeckung der Nachhaltigkeit: Kulturgeschichte eines Begriffs.* Munich: Antje Kunstmann.

Gross, C. (2011, 30 Mar). Taxi of tomorrow could reduce fleet fuel efficiency. *Gotham Gazette.* Retrieved March 2, 2012, from http://www.gothamgazette.com/article/searchlight/20110330/203/3501.

Gunder, M. (2006). Sustainability: Planning's saving grace or road to perdition? *Journal of Planning Education and Research*, 26(2), 208–21. doi:10.1177/0739456X06289359.

Häberli, M. and Schneider, N. (2002). *Nachhaltigkeitsvergleich Zürich - Berlin.* Opladen: Leske + Budrich.

Habermas, J. (1985). *The Theory of Communicative Action, Vol. 1: Reason and the Rationalization of Society.* Boston, MA: Beacon Press.

Hackenberg, K., Oostendorp, R., and Wiegandt, C-C. (2010). Schöne Städte durch große Pläne? – Baukultur und integrierte Stadtentwicklungsplanung. *Raumforschung und Raumordnung*, 68(6), 483–97. doi:10.1007/s13147-010-0059-x.

Hackworth, J. (2007). *The Neoliberal City: Governance, Ideology, and Development in American Urbanism.* Ithaca, NY: Cornell University Press.

Hahn, J. (2010). Creative cities and (un)sustainability – cultural perspectives. Cultura21 eBooks Series on Culture and Sustainability, Vol. 3. Retrieved March 28, 2015, from http://magazin.cultura21.de/_data/magazin-cultura21-de_addwp/2011/02/Julia-Hahn-c21-ebook-vol3.pdf.

Hajer, M.A. (1995). *The Politics of Environmental Discourse: Ecological Modernization and the Policy Process.* Oxford: Oxford University Press.

Harvey, D. (2005). *A Brief History of Neoliberalism.* Oxford: Oxford University Press.

Hawkins, C.V. and Wang, X. (2011). Sustainable development governance: Citizen participation and support networks in local sustainability initiatives. *Public Works Management & Policy*, 17(1), 7–29. doi:10.1177/1087724X11429045.

Healey, P. and Hillier, J. (1996). Communicative micropolitics: A story of claims and discourses. *International Planning Studies*, 1(2), 165–84.

Heiland, S., Tischer, M., Döring, T., Pahl, T., and Jessel, B. (2003). *Indikatoren zur Zielkonkretisierung und Erfolgskontrolle im Rahmen der Lokalen Agenda 21.* Umweltbundesamt No. UBA-FB 000513. Berlin. Retrieved September 10, 2010, from http://www.umweltdaten.de/publikationen/fpdf-l/2400.pdf.

Heinberg, R. (2011). *The End of Growth: Adapting to our New Economic Reality*. Retrieved draft form July 20, 2011, from http://www.postcarbon.org/article/178709-the-end-of-growth.

Heinelt, H. (2006). Planung und Governance. Der Beitrag der Governance-Debatte zum Planungsverständnis. In: Selle, K. (Ed.): *Zur räumlichen Entwicklung beitragen. Konzepte. Theorien. Impulse*. Planung neu denken, Bd. 1. Dortmund, 235–47.

Holden, M. (2006). Sustainable Seattle: The case of the prototype sustainability indicators project. In M.J. Sirgy, D. Rahtz, and D. Swain (eds), *Community Quality-of-life Indicators: Best cases II* (pp. 177–202). Amsterdam: Springer.

Holden, M. (2007). Revisiting the local impact of community indicators projects: Sustainable Seattle as prophet in its own land. *Applied Research in Quality of Life, 1*, 253–77. doi: 10.1007/s11482-007-9020-8.

Holden, M. (2008). Social learning in planning: Seattle's sustainable development codebooks. *Progress in Planning*, 69(1), 1–40. doi:10.1016/j.progress.2007.12.001.

Holden, M. (2011). Public participation and local sustainability: Questioning a common agenda in urban governance. *International Journal of Urban and Regional Research*, 35(2), 312–29. doi:10.1111/j.1468-2427.2010.00957.x.

Holgersen, S. (2014). *The Rise (and Fall?) of Post-industrial Malmö: Investigations of City-crisis Dialectics*. Published doctoral dissertation. Lund, Sweden: Lund University. ISBN: 978-91-7473-888-9.

Holman, N. (2009). Incorporating local sustainability indicators into structures of local governance: A review of the literature. *Local Environment*, 14(4), 365–75. doi: 10.1080/13549830902783043.

Hunter College Department of Urban Affairs & Planning. (n.d.) *Sustainability Watch*. Hunter College website. Retrieved May 14, 2014, from http://www.hunter.cuny.edu/ccpd/sustainability-watch.

ICLEI—Local Governments for Sustainability USA, and The Mayor's Office of Long-Term Planning and Sustainability, City of New York. (2010). *The Process Behind PlaNYC: How the City of New York Developed its Comprehensive Long-term Sustainability Plan*. Case study report. Retrieved May 14, 2011, from http://nytelecom.vo.llnwd.net/o15/agencies/planyc2030/pdf/iclei_planyc_case_study_201004.pdf.

IFEU and Stadt Heidelberg. (1997). *Nachhaltiges Heidelberg: Für eine lebenswerte UmWelt. Darstellung und Bewertung bisheriger Aktivitäten der Stadtverwaltung und Vorschläge für eine lokale Agenda 21*. Heidelberg. Schmidt, M., editor.

Innes, J.E. (1990). *Knowledge and Public Policy: The Search for Meaningful Indicators*. New Brunswick, NJ: Transaction Publishers.

Innes, J.E. (1995). Planning theory's emerging paradigm: Communicative action and interactive practice. *Journal of Planning Education and Research*, 14(3), 183–9. doi:10.1177/0739456X9501400307.

Innes, J.E. and Booher, D.E. (2000). Indicators for sustainable communities: A strategy building on complexity theory and distributed intelligence. *Planning Theory & Practice*, 1(2), 173–86. doi:10.1080/14649350020008378.

Innes, J.E. and Booher, D.E. (2010). *Planning with Complexity: An Introduction to Collaborative Rationality for Public Policy*. London: Routledge. [Kindle edition].

International Living Future Institute. (2013, 30 Dec). Miller Hull Partnership's Bullitt Center wins WAN's Sustainable Building of the Year award. International Living Future Institute website. Retrieved May 2, 2014, from http://living-future.org/news/miller-hull-partnerships-bullitt-center-wins-wans-sustainable-building-year-award.

Jackson, T. (2009). *Prosperity Without Growth? The Transition to a Sustainable Economy*. Report to the UK Sustainable Development Commission. Retrieved April 9, 2011, from http://www.sd-commission.org.uk/publications.php?id=914.

Jepson, E.J., Jr. (2004). The adoption of sustainable development policies and techniques in U.S. cities: How wide, how deep, and what role for planners? *Journal of Planning Education and Research*, 23(3), 229–41. doi:10.1177/0739456X03258638.

Jepson, E.J. and Edwards, M.M. (2010). How possible is sustainable urban development? An analysis of planners' perceptions about New Urbanism, smart growth and the ecological city. *Planning Practice and Research*, 25(4), 417–37. doi:10.1080/02697459.2010.511016.

Jerolleman, A. (2013). *The Privatization of Hazard Mitigation: A Case Study of the Creation and Implementation of a Federal Program*. Unpublished doctoral dissertation, University of New Orleans.

Johnson, S.R. (2004). The myth and reality of Portland's engaged citizenry and process-oriented government. In: Ozawa, C.P., Ed. *The Portland edge: Challenges and Successes in Growing Communities*. Washington, DC: Island Press, 102–17.

Jonas, A.E.G. and While, A. Greening the entrepreneurial city?: Looking for spaces of sustainability politics in the competitive city. In: Krueger, R. and Gibbs, D., eds. *The Sustainable Development Paradox: Urban Political Economy in the United States and Europe*. New York: The Guilford Press, 123–59.

Jordan, A. (2008). The governance of sustainable development: Taking stock and looking forwards. *Environment and Planning C: Government and Policy*, 26(1), 17–33. doi:10.1068/cav6.

Jörgensen, K. (2012). Governance for sustainable development in the German Bundesländer. In H. Bruyninckx, S. Happaerts, and K. van den Brande (eds.), *Sustainable Development and Subnational Governments: Policy-making and Multi-Level Interactions*. New York: Palgrave Macmillan, 103−119.

Journel, C.M., Duchene, F., Coanus, T., and Martinais, E. (2003). Devising local sustainable development indicators: From technical issues to bureaucratic stakes. The Greater Lyons experience. *Local Environment*, 8(6), 615–26. doi:10.1080/1354983032000152725.

Kahan, D., Peters, E., Dawson, E.C., and Slovic, P. (2013). Motivated numeracy and enlightened self-government. *Yale Law School, Public Law Working Paper No. 307*. Available at SSRN: http://ssrn.com/abstract=2319992.

Kallis, G. (2011). In defence of degrowth. *Ecological Economics*, 70(5), 873–80. Elsevier B.V. doi: 10.1016/j.ecolecon.2010.12.007.

Kamieniecki, S. and Sanasarian, E. (1987). Conducting comparative research on environmental policy. *Natural Resources Journal*, 30(2), 321–39.

Karlenzig, W., with Marquardt, F., White, P., Yaseen, R., and Young, R. (2007). *How Green is your City? The SustainLane US City Rankings*. Gabriola Island, BC: New Society Publishers.

Katz, A. and Baron, E. (2011, May). Participatory planning: The sustainability plan's missing piece. *Gotham Gazette*. Accessed January 16, 2012, at http://www.gothamgazette.com/article/landuse/20110511/12/3525.

Keil, R. (2007). Sustaining modernity, modernizing nature: The environmental crisis and the survival of capitalism. In: Krueger, R. and Gibbs, D., eds. *The*

Sustainable Development Paradox: Urban Political Economy in the United States and Europe. New York: The Guilford Press, 41–65.

Kemp, R., Parto, S., and Gibson, R.B. (2005). Governance for sustainable development: Moving from theory to practice. *International Journal of Sustainable Development*, 8(1/2), 12–30.

Kern, K., Koll, C., and Schophaus, M. (2007). The diffusion of Local Agenda 21 in Germany: Comparing the German federal states. *Environmental Politics*, 16(4), 604–24.

Kidd, C.V. (1992). The evolution of sustainability. *Journal of Agricultural and Environmental Ethics*, 5(1), 1–26.

Kissler, G.R., Fore, K.N., Jacobson, W.S., Kittredge, W.P., and Stewart, S.L. (1998). State strategic planning: Suggestions from the Oregon experience. *Public Administration Review*, 58(4), 353–60.

Köckler, H. (2005). *Zukunftsfähigkeit nach Maß: Kooperative Indikatorenentwicklung als Instrument regionaler Agenda-Prozesse.* Wiesbaden: VS Verlag für Sozialwissenschaften.

Kolleck, N. (2011). *Global Governance, Corporate Responsibility und die diskursive Macht multinationaler Unternehmen: Freiwillige Initiativen der Wirtschaft für eine nachhaltige Entwicklung?* Baden-Baden: Nomos.

Kooiman, J. (2003). *Governing as Governance.* London: Sage.

Kooshian, C. and Winkelman, S. (2011). *Growing Wealthier: Smart Growth, climate Change and Prosperity.* Washington, DC: Center for Clean Air Policy. Retrieved February 27, 2011, from http://www.growingwealthier.info/.

Kopfmüller, J., Brandl, V., Jörissen, J., Paetau, M., Banse, G., Coenen, R., and Grunwald, A., eds. (2001). *Nachhaltige Entwicklung integrativ betrachtet: Konstitutive Elemente, Regeln, Indikatoren.* Berlin: edition sigma.

Kraemer, R.A. and Schreurs, M.A. (2007). *Federalism and environmentalism in the United States and Germany* (AICGS Policy Report No. 31). Washington, DC: American Institute for Contemporary German Studies. Retrieved April 21, 2014, from http://www.ecologic.eu/sites/files/publication/2009/201-42-aicgs-pr-31.pdf.

Krätke, S. (2004). City of talents? Berlin's regional economy, socio-spatial fabric and "worst practice" urban governance. *International Journal of Urban and Regional Research*, 28(3), 511–29.

Kroh, E. (2011, 13 Jul). The nitty-gritty of going beyond GDP. *Remapping Debate* website. Retrieved April 4, 2014, from http://www.remappingdebate.org/article/nitty-gritty-going-beyond-gdp?page=0,4.

Krueger, R. and Gibbs, D. (2007). *The Sustainable Development Paradox: Urban Political Economy in the United States and Europe.* New York: The Guilford Press.

Kushner, J.A. (2003). *Comparative Urban Planning Law: An introduction to urban Land Development Law in the United States through the Lens Of Comparing the Experience of Other Nations.* Durham, NC: Carolina Academic Press.

Lafferty, W.M., ed. (2004). *Governance for Sustainable Development: The Challenge of Adapting form to Function.* Cheltenham: Edward Elgar.

Lafferty, W.M. and Meadowcroft, J., eds. (2001). *Implementing Sustainable Development: Strategies and Initiatives in High Consumption Societies.* Oxford: Oxford University Press.

LaFond, M.A. (1999). *From Century 21 to Local Agenda 21: Sustainable Development and Local Urban Communities in East and West Berlin (Germany), and Seattle (United States).* PhD dissertation for Interdisciplinary PhD in Urban Design and Planning. University of Washington.

Lang, A. (2003). *Ist Nachhaltigkeit messbar? Eine Gegenüberstellung von Indikatoren und Kriterien zur Bewertung nachhaltiger Entwicklung unter Berücksichtigung der Rahmenbedingungen in Deutschland und Frankreich. Überlegungen auf der Basis der Ergebnisse eines Forschungsvorhabens.* Stuttgart: ibidem-Verlag.

Larsen, C. (1996). What should be the leading principles of land use planning? A German perspective. *Vanderbilt Journal of Transnational Law*, 29, 967–1021.

Laschever, E.S. (1998). An overview of Washington's Growth Management Act. *Pacific Rim Law & Policy Journal*, 7(3), 657–77.

Latouche (2010). *Farewell to Growth.* Cambridge: Polity Press.

Leukhardt, F. and Allen, S. (2012). How environmentally focused is the German sustainability strategy? A critical discussion of the indicators used to measure sustainable development in Germany. *Environment, Development and Sustainability*, 15(1), 149–66. doi:10.1007/s10668-012-9380-6.

LIAISE. (2010). *Sustainability Assessment.* LIAISE Innovation Report No. 1. Retrieved October 11, 2013, from http://www.liaise-noe.eu/system/files/LIAISE_innovation_report_01.pdf.

Light, M.A. (1999). Different ideas of the city: Origins of metropolitan land-use regimes in the United States, Germany, and Switzerland. *The Yale Journal of International Law*, 24, 577–611.

Lindblom, C.E. (1990). *Inquiry andChange: The Troubled Attempt to Understand and Shape Society.* New Haven, CT: Yale University Press.

Lindblom, C.E. and Cohen, D.K. (1979). *Usable Knowledge: Social Science and Social Problem Solving.* New Haven, CT: Yale University Press.

Lloyd, B.D. (2000). Accommodating growth or enabling sprawl? The role of population growth projections in comprehensive planning under the Washington state Growth Management Act. *Gonzaga Law Review*, 36(1), 73–151.

Luhde-Thompson, N. (2004). Governing sustainable cities. *Local Environment*, 9(5), 481–5. doi:10.1080/1354983042000264919.

Luks, F., Ed. (2007). *Die Deutung von Nachhaltigkeit in Wissenschaft und Politik und ihre Implikationen für die europäische Stadt- und Regionalentwicklung.* NEDS Working Paper No. 7. Hamburg. Retrieved February 11, 2013, from http://www.neds-projekt.de/Download/NEDS_WP_7_06_2007.pdf.

Luks, F. (2010). *Endlich im Endlichen, oder: Warum die Rettung der Welt Ironie und Großzügigkeit erfordert.* Marburg: Metropolis.

Magee, L. and Scerri, A. (2012). From issues to indicators: Developing robust community sustainability measures. *Local Environment*, 17(8), 915–33.

Maiello, A., Christovão, A.C., Nogueira de Paiva Britto, A.L., and Frey, M. (2013). Public participation for urban sustainability: Investigating relations among citizens, the environment and institutions – an ethnographic study. *Local Environment*, 18(2), 167–83.

Martinez, A. and Heim, K. (2012, 5 Apr). Amazon a virtual no-show in hometown philanthropy: The world's biggest online retailer is a minor player—at best—in local charitable giving. *The Seattle Times.* Retrieved April 14, 2012, from http://seattletimes.nwsource.com/html/businesstechnology/2017883663_amazonmain25.html.

McCormick, K., Anderberg, S., Coenen, L., and Neij, L. (2013). Advancing sustainable urban transformation. *Journal of Cleaner Production, 50*, 1–11. doi:10.1016/j.jclepro.2013.01.003.

McGann, J.G. and Johnson, E.C. (2005). *Comparative Think Tanks, Politics and Public Policy.* Cheltenham, UK: Edward Elgar.

McKeown, T.J. (2004). Case studies and the limits of the quantitative worldview. In: Brady, H.E. and Collier, D., eds. *Rethinking Social Inquiry: Diverse Tools, Shared Standards.* New York: Rowman & Littlefield, 139–67.

McLaren, C. (2012, 16 May). The history behind Berlin's cheap rents. *Next City* blog. Retrieved April 15, 2014, from http://nextcity.org/daily/entry/the-history-behind-berlins-cheap-rents.

Meadowcroft, J. (1997). Planning, democracy and the challenge of sustainable development. *International Political Science Review*, 18(2), 167–89. doi:10.1177/019251297018002004.

Meadowcroft, J. (2004). Participation and sustainable development: Modes of citizen, community and organisational involvement. In: Lafferty, W.M., ed. *Governance for Sustainable Development: The Challenge of Adapting Form to Function.* Cheltenham: Edward Elgar, 162–90.

Meadowcroft, J., Farrell, K.N., and Spangenberg, J. (2005). Developing a framework for sustainability governance in the European Union. *International Journal of Sustainable Development*, 8(1/2), 3–11.

Meadows, D. (1998). *Indicators and Information Systems for Sustainable Development.* A report to the Balaton Group. Hartland, VT: The Sustainability Institute.

Meadows, D. (2008). *Thinking in Systems: A Primer.* White River Junction, VT: Chelsea Green. Kindle edition.

Meadows, D.H., Meadows, D.L., Randers, J., and Behrens, W.W. (1972). *The Limits to Growth.* New York: New American Library.

Meadows, D.H., Randers, J., and Meadows, D.L. (2004). *The Limits to Growth: The 30-year Update.* White River Junction, Vt.: Chelsea Green. Kindle edition.

Menzel, H-J. (2000). Mehr Präambel und Projekte als Programme und Prozeß. In: GERMANWATCH Regionalgruppe Hamburg, eds. (2000). *Agenda 21:*

Hamburgs mühsamer Weg ins 21. Jahrhundert. Dritter Nord-Süd WatchBericht. Hamburg: Konkret Literatur Verlag, 54–61.
Menzel, H-J. (2004). Wachsende Stadt – Nachhaltige Stadt. In: Altrock, U. and Schubert, D., eds. *Wachsende Stadt: Leitbild – Utopie – Vision?* Wiesbaden: VS Verlag. 57–66.
Menzel, H-J. (2008). Macht und Ohnmacht zivilgesellschaftlicher Akteure in der Hamburger Nachhaltigkeitspolitik. In: Bauriedl, S., Schindler, D., and Winkler, M., eds. *Stadtzukünfte denken: Nachhaltigkeit in europäischen Stadtregionen.* Munich: oekom Verlag, 158–78.
Metro. (2011). *Summary of 2040 Growth Concept.* Portland, OR: Metro.
Miller, D. (2004). Decentralized planning for sustainable urban development: The Washington State Growth Management Act and its application in the Seattle metropolitan area. In: Feitelson, E., ed. *Advancing Sustainability at the Sub-national Level: The Potential and Limitations Of Planning.* Aldershot: Ashgate, 45–56.
Miller, T. (2011). *Constructing Sustainability: A Study of Emerging Scientific Research Trajectories.* Published doctoral dissertation. Arizona State University, Tempe, AZ. Retrieved March 16, 2014, from http://repository.asu.edu/attachments/56608/content/Miller_asu_0010E_10655.pdf.
Misaras, L.R. (2007). *Oregon State Legislators' Use of Oregon Benchmark Data in Legislative Decision-making.* MPA thesis. Retrieved May 24, 2010, from http://www.oregon.gov/DAS/OPB/docs/Committees/thesis_draft_x.doc.pdf.
Mishan (1993 [1967]). *The Costs of Economic Growth.* Westport, CT: Praeger Publishers.
Mitra, A. (2003). *Painting the Town Green: The Use of Urban Sustainability Indicators in the United States of America.* London: RICS Foundation. Retrieved May 13, 2010, from http://www.rics.org/site/download_feed.aspx?fileID=2947&fileExtension=PDF.
Molotch, H. (1976). The city as a growth machine: Toward a political economy of place. *American Journal of Sociology*, 82(2), 309–32. doi:10.1086/226311.
Monstadt, J. (2007). Urban governance and the transition of energy systems: Institutional change and shifting energy and climate policies in Berlin. *International Journal of Urban and Regional Research*, 31(2), 326–43. doi:10.1111/j.1468-2427.2007.00725.x.
Moore, J., Kissinger, M., and Rees, W.E. (2013). An urban metabolism and ecological footprint assessment of Metro Vancouver. *Journal of Environmental Management*, 124, 51–61. doi:10.1016/j.jenvman.2013.03.009.
Mori, K. and Christodoulou, A. (2011). Review of sustainability indices and indicators: Towards a new City Sustainability Index (CSI). *Environmental Impact Assessment Review*, 4–6. doi:10.1016/j.eiar.2011.06.001.
Morris, C. and Pehnt, M. (2012). *Energy Transition: The German Energiewende.* Heinrich Böll Foundation. Retrieved October 15, 2013, from http://www.energytransition.de.

Moss, M.L. and Grunkemeyer, W.T. (2010). Building shared visions for sustainable communities. *Community Development*, 41(2), 240–54. doi:10.1080/15575330903477309.

Myers, N. and Simon, J.L. (1994). *Scarcity or Abundance? A Debate on the Environment.* New York: W.W. Norton & Company.

Naparstek, A. (2006, 21 Sep). Bloomberg sustainability announcement. *Streetsblog NYC* website. Retrieved April 9, 2014, from http://www.streetsblog.org/2006/09/21/bloomberg-sustainability-announcement/.

Newman, L. and Dale, A. (2005). The role of agency in sustainable local community development. *Local Environment*, 10(5), 477–86. doi:10.1080/13549830500203121.

Niestroy, I. (2005). Individual country analysis – Germany. *Sustaining Sustainability: A Benchmark Study on National Strategies Towards Sustainable Development and the Impact of Councils in Nine EU Member States*. EEAC series Background study no. 2. The Hague. Retrieved January 20, 2011, from http://www.eeac-net.org/download/EEAC_3Germany.pdf.

Northwest Environment Watch. (2004). *Cascadia Scorecard: Seven Key Trends Shaping the Northwest*. Seattle, WA. Retrieved April 6, 2012, from http://www.sightline.org/download/2209.

Oates, D. (2006). *City Limits: Walking Portland's Boundary.* Corvallis, OR: Oregon State University Press.

Office of City Auditor. (2007). *Neighborhood Plan Implementation.* City of Seattle. Retrieved May 13, 2014, from http://www.seattle.gov/neighborhoods/documents/PublishedNPI.pdf.

O'Neill, J. (2009). Sustainability, welfare and value over time. In: Adger, W.N. and Jordan, A., eds. *Governing Sustainability.* Cambridge: Cambridge University Press, 283–304.

Ornstein, N. (2013, 12 Dec). Farm bill serves as poster child for Congressional dysfunction. *The National Journal.* Retrieved July 9, 2014, from http://www.nationaljournal.com/washington-inside-out/farm-bill-serves-as-poster-child-for-congressional-dysfunction-20131211.

Ott, K. and Döring, R. (2008). *Theorie und Praxis starker Nachhaltigkeit.* Marburg: Metropolis-Verlag.

Otto-Zimmermann, K. (2012). From Rio to Rio + 20: The changing role of local governments in the context of current global governance. *Local Environment*, 17(5), 511–16.

Owen, D. (2009). *Green Metropolis: Why Living Smaller, Living Closer, and Driving Less are the Keys to Sustainability.* New York: Riverhead Books.

Owen, D. (2011). *The Conundrum: How Scientific Innovation, Increased Efficiency, and Good Intentions Can Make Our Energy and Climate Problems Worse.* New York: Riverhead Books.

Ozawa, C.P., Ed. (2004). *The Portland Edge: Challenges and Successes in Growing Communities.* Washington, DC: Island Press.

Pallagst, K.M. (2007). *Growth Management in the US.* Aldershot: Ashgate.

Parker, R.A. (1997). A stealth urban policy in the US? Federal spending in five large metropolitan regions, 1984–93. *Urban Studies*, 34(11), 1831–50. doi:10.1080/0042098975277.

Parris, T.M. and Kates, R.W. (2003). Characterizing and measuring sustainable development. *Annual Review of Environment and Resources*, 28(1), 559–86. doi:10.1146/annurev.energy.28.050302.105551.

Partnership for Sustainable Communities (PSC). (2013). *About us.* Organization's website. Retrieved April 1, 2014, from http://www.sustainablecommunities.gov/aboutUs.html.

The Pastille Consortium. (2002). *Indicators into Action: Local Sustainability Indicator Sets in their Context.* Retrieved May 27, 2010, from http://www.lse.ac.uk/collections/PASTILLE/images/FinalReportWeb.pdf.

Patel, R. (2009). *The Value of Nothing: How to Reshape Market Society and Redefine Democracy.* New York: Picador.

Pearsall, H. and Pierce, J. (2010). Urban sustainability and environmental justice: Evaluating the linkages in public planning/policy discourse. *Local Environment*, 15(6), 569–80. doi:10.1080/13549839.2010.487528.

Peters, D. (2003). *Planning for a Sustainable Europe? A Case Study of EU Transport Infrastructure Investment Policy in the Context of Eastern Enlargement.* Unpublished PhD dissertation. Retrieved March 10, 2013, from http://www.geschundkunstgesch.tu-berlin.de/fileadmin/fg95/Hauptordner_Megaprojekte/Peters2003RutgersPhDPlanningforASustainableEurope.pdf.

Pettenger, M.E. (2007). Introduction: Power, knowledge and the social construction of climate change. In: Pettenger, M.E., ed. *The Social Construction of Climate Change: Power, Knowledge, Norms, Discourses.* Farnham: Ashgate, 1–19.

Pettibone, L. (2012). Sustainability indicators in an evolving sustainability information system: The case of Seattle. *Paper Presented at 2012 Berlin Conference on the Human Dimensions of Global Environmental Change*, Berlin, 5–6 October.

Pettibone, L. (2014a). *Incorporating Sustainability Principles in Six U.S. and German Cities: The Role of Strategic Plans, Sustainability indicators, and Sustainability-minded Groups.* Unpublished doctoral dissertation. Environmental Policy Research Centre: Freie Universität Berlin.

Pettibone, L. (2014b). Sustainability needs citizens, but do citizens need sustainability? Introduction to a disruptive worldview. *Open Citizenship* 5(1): 8–19. Available online at: opencitizenship.eu/ojs/index.php/opencitizenship/issue/current/showToc.

Pettibone, L. (2015). Introduction: The need for integrative and interdisciplinary approaches for urban sustainability. *Journal of Environmental Studies & Sciences* 5(2). DOI 10.1007/s13412-014-0211-y.

Pfister, T. and Flear, M.L. (2014). Sustainable technoscientific development and the innovator citizen – way out or deeper into the crisis? *Open Citizenship* 5(2), 28–39.

Pierce, J.C., Budd, W.W., and Lovrich, Jr., N.P. (2011). Resilience and sustainability in US urban areas. *Environmental Politics*, 20(4), 566–84.

Pinfield, G. (1997). The use of indicators in local sustainable development planning: A response to Jeb Brugmann. *Local Environment*, 2(2), 185–7.

Pinzler, P. (2013, 21 Feb). Was ist Lebensqualität? *Die Zeit.* Retrieved April 28, 2014, from http://www.zeit.de/2013/09/Enquete-Alternative-zum-Wachstum-Lebensqualitaet.

Pollard, III, O.A. (2000). Smart growth: The promise, politics, and potential pitfalls of emerging growth management strategies. *Virginia Environmental Law Journal* 19(3): 247–86. In: Kushner, J.A. (2003). *Comparative Urban Planning Law: An Introduction to Urban Land Development Law in the United States through the Lens of Comparing the Experience of Other Nations.* Durham, NC: Carolina Academic Press, 274–78. (abridged).

Portney, K.E. (2003). *Taking Sustainable Cities Seriously: Economic Development, the Environment, and Quality of Life in American Cities.* Cambridge, Mass.: MIT Press. Kindle edition.

President's Council on Sustainable Development. (1996). *Sustainable America: A New Consensus for the Prosperity, Opportunity and a Healthy Environment for the Future.* Washington, DC. Retrieved November 16, 2010, from http://clinton4.nara.gov/PCSD/Publications/TF_Reports/amer-top.html.

President's Council on Sustainable Development. (1997). *Building on Consensus: A Progress Report on Sustainable America.* Washington, DC. Retrieved November 16, 2010, from http://clinton4.nara.gov/PCSD/Publications/Progress_Report.html

President's Council on Sustainable Development. (1999). *Towards a sustainable America: Advancing Prosperity, Opportunity, and a Healthy Environment for the 21st Century.* Washington, DC. Retrieved November 16, 2010, from http://clinton4.nara.gov/media/pdf/tsa.pdf.

Putnam, R.D. and Feldstein, L.M., with Cohen, D. (2003). Portland: A positive epidemic of civic engagement. In: *Better Together: Restoring the American Community.* New York: Simon & Schuster, 241–68.

Rabe, B.G., Román, M., and Dobelis, A.N. (2005). State competition as a source driving climate change mitigation. *NYU Environmental Law Journal*, 14, 1–53.

Ragin, C.C. (2004). Turning the tables: How case-oriented research challenges variable-oriented research. In: Brady, H.E. and Collier, D., eds. *Rethinking Social Inquiry: Diverse Tools, Shared Standards.* New York: Rowman & Littlefield, 123–38.

Ralston, H. (2012). *State/Länder Environmental Partnerships In Global Governance for Sustainable Development.* Unpublished doctoral dissertation. Environmental Policy Research Centre. Freie Universität Berlin.

German Council for Sustainable Development [RNE]. (2013). *Sustainability – Made in Germany: The Second Review by a Group of International Peers, Commissioned by the German Federal Chancellery*. Berlin. Retrieved October

16, 2013, from http://www.nachhaltigkeitsrat.de/uploads/media/20130925_Peer_Review_Sustainability_Germany_2013_03.pdf.

Reed, M.S., Fraser, E.D.G., Morse, S., and Dougill, A.J. (2005). Integrating methods for developing sustainability indicators to facilitate learning and action. *Ecology and Society*, 10(1), r3. Retrieved May 10, 2010, from http://www.ecologyandsociety.org/vol10/iss1/resp3/.

Rees, W.E. (1997). Is "sustainable city" an oxymoron? *Local Environment*, 2(3), 303–10.

Reusswig, F., Hirschl, B., Lass, W., et al. (2014). *Machbarkeitsstudie Klimaneutrales Berlin 2050*. Berlin & Potsdam. Retrieved April 15, 2014, from http://www.stadtentwicklung.berlin.de/umwelt/klimaschutz/studie_klimaneutrales_berlin/download/Machbarkeitsstudie_Berlin2050_Hauptbericht.pdf.

Richardson, H.W. and Gordon, P. (2001). Sustainable Portland? A critique, and the Los Angeles counterpoint. *Paper presented at the ACSP conference*, Cleveland, Ohio, 10 November. Available from: http://www.ref.usc.edu/~pgordon/pdf/SUSTAINABLE_PORTLAND.pdf.

Rockström, J. (2009). A safe operating space for humanity. *Nature*, 461(24), 472–5.

Rosan, C.D. (2007). *Metropolitan Governance and Local Land Use Planning in Boston, Denver, and Portland*. Unpublished master's thesis. Massachusetts Institute of Technology.

Rosan, C.D. (2012). Can PlaNYC make New York City "greener and greater" for everyone?: Sustainability planning and the promise of environmental justice. *Local Environment*, 17(9), 959–76.

Rose-Ackerman, S. (1995). *Controlling Environmental Policy: The Limits of Public Law in Germany and the United States*. New Haven, CT: Yale University Press.

Rosenström, U. (2006). Exploring the policy use of sustainable development indicators: Interviews with Finnish politicians. *The Journal of Transdisciplinary Environmental Studies*, 5(1–2), 1–13.

Rubin, H.J. and Rubin, I.S. (2005). *Qualitative Interviewing: The Art of Hearing Data*. Second edition. Thousand Oaks, CA: Sage Publications. Kindle edition.

Rutland, T. and Aylett, A. (2008). The work of policy: Actor networks, governmentality, and local action on climate change in Portland, Oregon. *Environment and Planning D: Society and Space*, 26(4), 627–46. doi:10.1068/d6907.

Rybczynski, W. (2007). *Last Harvest. How a Cornfield became New Daleville: Real Estate Development in America from George Washington to the Builders of the Twenty-first century, and Why we Live in Houses Anyway*. New York: Scribner.

Rydin, Y., Holman, N. and Wolff, E. (2003). Local sustainability indicators. *Local Environment*, 8(6), 581–9. doi:10.1080/1354983032000152707.

Sabatier, P.A. and Jenkins-Smith, H.C., eds. (1993). *Policy change and learning: An Advocacy Coalition Approach*. Boulder, CO: Westview Press.

Sachverständigenrat für Umweltfragen [SRU]. (2012). *Umweltgutachten 2012: Verantwortung in einer begrenzten Welt*. Berlin: Erich Schmidt Verlag.

Retrieved April 23, 2014, from http://www.umweltrat.de/SharedDocs/Downloads/DE/01_Umweltgutachten/2012_06_04_Umweltgutachten_HD.pdf?_blob=publicationFile.

Saha, D. (2009). Empirical research on local government sustainability efforts in the USA: Gaps in the current literature. *Local Environment*, 14(1), 17–30. doi: 10.1080/13549830802522418.

Sanders, J.C. (2010). *Seattle and the Roots of Urban Sustainability: Inventing Ecotopia.* Pittsburgh, PA: University of Pittsburgh Press.

Sandilya, H. and Danna, R. (2013). Cultivating citizenship: The impact of urban agriculture. *Open Citizenship*, 4(2), 54–64.

Sawicki, D.S. (2002). Improving community indicator systems: Injecting more social science into the folk movement. *Planning Theory & Practice*, 3(1), 13–32. doi: 10.1080/14649350220117780.

Schindler, D. (2008). Die Magie der Zahlen. Nachhaltigkeitsindikatoren im Wettbewerb der Städte und Regionen – das Beispiel Hamburg. In: Bauriedl, S., Schindler, D., and Winkler, M., eds. *Stadtzukünfte denken: Nachhaltigkeit in europäischen Stadtregionen.* Munich: oekom Verlag, 179–99.

Schindler, D. (2011). *Urban Governance: Wandel durch das Leitbild Nachhaltigkeit?* Kassel: Kassel University Press.

Schmidthals, M. and Kliche, F. (2004). *Berliner Nachhaltigkeitsindex: Indikatoren für die Lokale Agenda 21*. Berlin: Unabhängiges Institut für Umweltfragen.

Scholl, B., Elgendy, H., and Nollert, M. (2007). *Raumplanung in Deutschland – formeller Aufbau und zukünftige Aufgaben; Spatial Planning in Germany – formal Structure and Future Tasks* [in German and English]. Schriftenreihe des Instituts für Städtebau und Landesplanung, Universität Karlsruhe (TH), Band 35. Karlsruhe: Universitätsverlag Karlsruhe. Retrieved July 20, 2013, from http://digbib.ubka.uni-karlsruhe.de/volltexte/1000006670.

Schreurs, M.A. (2003). *Environmental Politics in Japan, Germany, and the United States.* Cambridge: Cambridge University Press.

Schreurs, M.A. (2007). Federalism and environmentalism in the United States. In: Kraemer, R.A. and Schreurs, M.A. (2007). *Federalism and Environmentalism in the United States and Germany* (AICGS Policy Report No. 31). Washington, DC: American Institute for Contemporary German Studies. 35–48. Retrieved April 21, 2014, from http://www.ecologic.eu/sites/files/publication/2009/201-42-aicgs-pr-31.pdf.

Schreurs, M.A., Selin, H., and VanDeveer, S. (eds). (2009). *Transatlantic Environment and Energy Politics: Comparative and International Perspectives.* Farnham: Ashgate.

Schreurs, M.A., Selin, H., and VanDeveer, S. (2009a). "Expanding transatlantic relations: Implications for environment and energy politics." In Schreurs, M.A., Selin, H., and VanDeveer, S. (eds), *Transatlantic Environment and Energy Politics: Comparative and International Perspectives.* Farnham: Ashgate, 1–18.

Schröder, D. (2013, 13 Oct). Das Tor zur Provinz: Der Hamburger Hafen droht im internationalen Wettbewerb um Riesenfrachter bedeutungslos zu werden. *Zeit Online*. Retrieved April 15, 2014, from http://www.zeit.de/2013/41/hamburg-hafen-containerschiff-elbvertiefung.

Schwengler, C. (2014). Understanding urban sustainability through newspaper discourse: A look at Germany. *Journal of Environmental Studies & Sciences* 5(2). doi: 10.1007/s13412-014-0188-6.

Seattle City Light. (n.d.). *A Brief History*. Website. Retrieved May 14, 2014, from http://www.seattle.gov/light/aboutus/history/ab5_brhs.htm.

Selle, K., with Zalas, L., ed. (2006). *Zur räumlichen Entwicklung beitragen: Konzepte. Theorien. Impulse.* Planung neu denken, Edition Stadtentwicklung, Band 1. Dortmund: Verlag Dorothea Rohn.

Selman, P. and Parker, J. (1997). Citizenship, civicness and social capital in Local Agenda 21. *Local Environment*, 2(2), 171–84.

Seltzer, E. (2004). It's not an experiment: Regional planning at Metro, 1990 to the present. In: Ozawa, C.P., Ed. *The Portland Edge: Challenges and Successes in Growing Communities.* Washington, DC: Island Press, 35–60.

Senatsverwaltung für Stadtentwicklung und Umwelt [SSU]. (2012). *Berliner Strategie zur Biologischen Vielfalt: Begründung, Themenfelder und Ziele.* Berlin. Retrieved March 13, 2015, from http://www.stadtentwicklung.berlin.de/natur_gruen/naturschutz/downloads/publikationen/biologische_vielfalt_strategie.pdf.

Senatsverwaltung für Stadtentwicklung und Umwelt [SSU]. (2013). *Stadtentwicklungskonzept Berlin 2030: Statusbericht.* Berlin. Retrieved May 2, 2013, from http://www.stadtentwicklung.berlin.de/planen/stadtentwicklungskonzept/download/status/2013-04-08_Statusbestimmung_StEKBerlin.pdf.

Senatsverwaltung für Stadtentwicklung und Umwelt [SSU]. (2014a). *Die BerlinStrategie – Stadtentwicklungskonzept Berlin 2030. Wie soll Berlin im Jahr 2030 aussehen?* Berlin. Retrieved May 7, 2014, from http://www.stadtentwicklung.berlin.de/planen/stadtentwicklungskonzept/download/strategie/StEK2030_BerlinStrategie_kompakt.pdf.

Senatsverwaltung für Stadtentwicklung und Umwelt [SSU]. (2014b). *BerlinStrategie: Stadtentwicklungskonzept Berlin 2030.* Berlin. Retrieved March 6, 2015, from http://www.stadtentwicklung.berlin.de/planen/stadtentwicklungskonzept/download/strategie/BerlinStrategie.pdf.

Sheehan, M. O'M, (Ed.). (2001). *City Limits: Putting the brakes on sprawl.* Worldwatch Paper No. 156. Retrieved February 29, 2012, from http://www.worldwatch.org/system/files/EWP156.pdf.

Siemens. (2011a). *German Green City Index: Analyse der Leistungen zwölf deutscher Großstädte im Bereich Umwelt- und Klimaschutz.* München. Retrieved February 17, 2012, from http://www.siemens.com/entry/cc/en/greencityindex.htm.

Siemens. (2011b). *The Green City Index: A Summary of the Green City Index Research Series.* Munich. Retrieved January 24, 2013, from http://www.siemens.com/entry/cc/de/greencityindex.htm.

Siemens. (2011c). *US and Canada Green City Index: Assessing the environmental Performance of 27 Major US and Canadian Cities.* Munich. Retrieved February 17, 2012, from http://www.siemens.com/entry/cc/en/greencityindex.htm.

Skocpol, T. (2013). *Naming the Problem: What it will take to counter extremism and engage Americans in the fight against global warming.* Paper presented at *Symposium on the Politics of America's Fight Against Global Warming.* Cambridge, MA, 14 February. Retrieved April 23, 2013, from http://www.scholarsstrategynetwork.org/sites/default/files/skocpol_captrade_report_january_2013y.pdf.

Smart Growth America. (2013). *Federal Involvement in Real Estate: A Call for Examination.* Washington, DC. Retrieved January 18, 2013, from http://www.smartgrowthamerica.org/documents/federal-involvement-in-real-estate.pdf.

Smedby, N. and Neij, L. (2013). Experiences in urban governance for sustainability: The Constructive Dialogue in Swedish municipalities. *Journal of Cleaner Production*, 50, 148–58. doi:10.1016/j.jclepro.2012.11.044.

Smith, R. and Wiek, A. (2012). Achievements and opportunities in initiating governance for urban sustainability. *Environment and Planning C: Government and Policy*, 30(3), 429–47. doi:10.1068/c10158.

Solow, R.M. (1956). A contribution to the theory of economic growth. *Quarterly Journal of Economics*, 70(1), 65–94.

SoundTransit. (2007). SoundTransit, Chronology. Retrieved January 12, 2012, from http://www.soundtransit.org/Documents/pdf/about/Chronology.pdf.

Stadt Heidelberg. (1995 Sept). *Solidarische Stadt. Dokumentation der Auftaktveranstaltung vom 8. Oktober 1994.* Heidelberg: City of Heidelberg.

Stadt Heidelberg. (1997). *Materialband: Synopse der Stellungnahmen zum Diskussionsentwurf und Skizzen zur Stadtentwicklung Heidelberg der Architektenkammergruppe Heidelberg.* Schriften zur Stadtentwicklung. Heidelberg: City of Heidelberg.

Stadt Heidelberg. (2002). *Wo stehen wir, was haben wir erreicht? Erster Bericht zur Umsetzung des Stadtentwicklungsplans Heidelberg 2010.* Heidelberg. Retrieved May 14, 2011, from http://www.heidelberg.de/servlet/PB/menu/1116400_l2/index.html [Also available in English as: *Where do we stand, what are our Achievements? A First Report on the Implementation of the Heidelberg City Development Plan 2010*].

Stadt Heidelberg. (2002 [1997]). *Stadtentwicklungsplan Heidelberg 2010: Leitlinien und Ziele.* Heidelberg. Retrieved May 14, 2011, from: http://www.heidelberg.de/servlet/PB/menu/1116400_l2/index.html [Also available in English as *Heidelberg City Development Plan 2010: Guidelines and goals*].

Stadt Heidelberg and GEWOS Institut. (2013). *Wohnraumsbedarfanalyse Heidelberg 2030.* Retrieved April 15, 2014, from http://www.heidelberg.

de/site/Heidelberg_ROOT/get/documents/heidelberg/Objektdatenbank/12/PDF/12_pdf_WohnraumbedarfsanalyseHD2030.pdf.
Stake, R.E. (1995). *The Art of Case Study Research.* Thousand Oaks, CA: SAGE Publications, Inc.
Staley, S.R. (2006). Sustainable development in American planning: A critical appraisal. *The Town Planning Review*, 77(1), 99–126.
Statistisches Amt Berlin-Brandenburg. (2010). *Berlin in Zahlen 2010.* Retrieved April 14, 2014, from https://www.statistik-berlin-brandenburg.de/produkte/faltblatt_brochure/berlin_in_Zahlen_de.pdf.
Statistisches Amt Berlin-Brandenburg. (2010b). Bevölkerungsentwicklung Berlin. *Zeitschrift für amtliche Statistik Berlin Brandenburg 3/2010.* Retrieved April 15, 2014, from https://www.statistik-berlin-brandenburg.de/Publikationen/Aufsaetze/2010/HZ_201003-01.pdf.
Statistisches Amt für Hamburg und Schleswig-Holstein. (2014). *Hochbautätigkeit in Hamburg 2012 und Wohnungsbestand in Hamburg 2012, 2011, 2010.* Retrieved April 15, 2014, from http://www.statistik-nord.de/uploads/tx_standocuments/F_II_1_2_4_j12_HH_neu.pdf.
Statistiches Bundesamt. (2011). *Daten aus dem Gemeindeverzeichnis Gerichtsbarkeit nach Fläche und Bevölkerung.* Wiesbaden: Statistisches Bundesamt. Retrieved January 29, 2014, from http://www.statistik-portal.de/Statistik-Portal/ISS/gerichtsnummern.pdf.
Statistisches Landesamt Baden-Württemberg. (2012). *Zukunft Baden-Württemberg: Indikatoren im Vergleich.* Stuttgart: Land Baden-Württemberg.
Stedman Jones, D. (2012). *Masters of the Universe: Hayek, Friedman, and the birth of Neoliberal Politics.* Princeton, NJ: Princeton University Press.
Steffen, W., Richardson, K., Rockström, J. et al. (2015). Planetary boundaries: Guiding human development on a changing planet. *Science.* DOI:10.1126/science.1259855.
Stigson, B., Babu, S.P., Bordewijk, J., O'Donnell, P., Haavisto, P., Morgan, J., and Osborn, D. (2009). *Peer Review on Sustainable Development Policies in Germany.* German Council for Sustainable Development. Retrieved January 18, 2013, from http://www.nachhaltigkeitsrat.de/uploads/media/RNE_Peer_Review_Report_November_2009_03.pdf.
Stirling, A. (2009). Participation, precaution and reflexive governance for sustainable development. In: Adger, W.N. and Jordan, A., eds. *Governing Sustainability.* Cambridge: Cambridge University Press, 193–225.
Stockdale, J. (1989). Pro-growth, limits to growth, and a sustainable development synthesis. *Society & Natural Resources*, *2*, 163–76.
Stone, D. (1996). *Capturing the Political Imagination: Think Tanks and the Policy Process.* London: Frank Cass.
Sustainable Seattle. (1998). *Indicators of Sustainable Community.* Seattle, WA. Retrieved April 30, 2010, from http://sustainableseattle.org/Programs/RegionalIndicators/1998IndicatorsRpt.pdf.

Svoboda, E., with Mika, E., and Berhie, S. (2008, 8 Feb). America's 50 greenest cities. *Popular Science* website. Retrieved February 13, 2014, from http://www.popsci.com/environment/article/2008-02/americas-50-greenest-cities.

Teichert, V. (2000). Überblick über Indikatorensysteme nachhaltiger Entwicklung auf kommunaler, regionaler und nationaler Ebene. In: Gehrlein, U., ed. *Wege zur Zukunftsbeständigkeit: Strategien und Instrumente zur Umsetzung des Leitbildes nachhaltiger Entwicklung.* Darmstädter interdisziplinäre Beiträge 3. Münster: agenda Verlag, 155–76.

Thompson, C. (2009, 17 Jul). The 15 most sustainable U.S. cities. *Grist.* Retrieved February 13, 2014, from http://grist.org/article/2009-07-16-sustainable-green-us-cities/.

Trapenberg Frick, K., Weinzimmer, D., and Waddell, P. (2014). The politics of sustainable development opposition: State legislative efforts to stop the United Nation's Agenda 21 in the United States. *Urban Studies*. [OnlineFirst edition.] doi:10.1177/0042098014528397.

Tryens, J. (2004). Using indicators to engage citizens: The Oregon Progress Board experience. *Statistics, Knowledge and Policy: OECD World Forum on Key Indicators.* Palermo, 10–13 November.

United Nations Conference on Environment and Development (UNCED). (1992). *Agenda 21 -- Action Plan for the Next Century* Rio de Janeiro: UNCED.

United Nations Conference on the Human Environment. (1972). *Declaration of the United Nations Conference on the Human Environment.* Retrieved March 6, 2013, from http://www.unep.org/Documents.Multilingual/Default.Print.asp?documentid=97&articleid=1503.

U.S. Census Bureau. (2012). *State and County Quickfacts*. Retrieved January 29, 2014, from http://quickfacts.census.gov/qfd/index.html# .

Van Begin, G. (2004). Inspiring futures for European local governments! *Local Environment*, 9(3), 213–16. doi:10.1080/1354983042000219333.

van den Bergh, J.C.J.M. (2011). Environment versus growth—A criticism of "degrowth" and a plea for "a-growth." *Ecological Economics*, 70(5), 881–90. Elsevier B.V. doi: 10.1016/j.ecolecon.2010.09.035.

Van Huijstee, M.M., Francken, M., and Leroy, P. (2007). Partnerships for sustainable development: A review of current literature. *Environmental Sciences*, 4(2), 75–89. doi:10.1080/15693430701526336.

van Zeijl-Rozema, A., Cörvers, R., Kemp, R., and Martens, P. (2008). Governance for sustainable development: A framework. *Sustainable Development*, 16(6), 410–21.

Venzke, A. (2011). *Berlin, Berlin – Geschichte einer Nation.* Würzburg: Arena.

Victor, D.G. (2006). Recovering sustainable development. *Foreign Affairs*, 85(1), 91–103.

VisionPDX. (2008). *Portland 2030: A Vision for the Future by the People of Portland, Oregon.* Portland. Retrieved July 23, 2013, from http://www.visionpdx.com/downloads/final%20vision%20document_Feb.pdf.

von Beust, O. (2004). Das Leitbild 'Metropole Hamburg – Wachsende Stadt.' In Altrock, U. and Schubert, D. (eds) *Wachsende Stadt: Leitbild – Utopie – Vision?* Wiesbaden: VS Verlag, 23–37.

Vonkeman, G.H., Ed. (2000). *Sustainable Development of European Cities and Regions*. Dordecht, Netherlands: Kluwer Academic Publishing.

Voß, J-P., Bauknecht, D., and Kemp, R. (eds). (2006). *Reflexive Governance for Sustainable Development*. Cheltenham: Edward Elgar.

Voß, J-P. and Kemp, R. (2006). Sustainability and reflexive governance: Introduction. In: Voß, J-P., Bauknecht, D., and Kemp, R., eds. *Reflexive Governance for Sustainable Development*. Cheltenham: Edward Elgar, 3–28.

Wälti, S. (2009). Governance for sustainable development: The United States and the European Union compared. In Schreurs, M.A., Selin, H., and VanDeveer, S. (eds), *Transatlantic Environment and Energy Politics: Comparative and International Perspectives*. Farnham: Ashgate, 41−54.

German Advisory Council on Global Change (WBGU). (2005). *Fighting Poverty through Environmental Policy. Flagship Report 2004*. Berlin, Heidelberg, New York: Springer. fileadmin/templates/dateien/veroeffentlichungen/hauptgutachten/jg2011/wbgu_jg2011_en.pdf.

German Advisory Council on Global Change (WBGU). (2011). *World in Transition: A Social Contract for Sustainability*. Flagship report. Berlin. Retrieved October 11, 2012, from http://www.wbgu.de/fileadmin/templates/dateien/veroeffentlichungen/hauptgutachten/jg2011/wbgu_jg2011_en.pdf.

Weiss, C.H. (1979). The many meanings of research utilization. *Public Administration Review*, 426–31.

Weiss, C.H. and Bucuvalas, M.J. (1980). Truth tests and utility tests: Decision-makers' frames of reference for social science research. *American Sociological Review*, 45(2), 302–13.

Weiss, C. (1983). Ideology, interest, and information: The basis of policy decisions. In Callahan, D. and Jennings, B. (eds). *Ethics, the Social Sciences, and Policy Analysis*. New York: Plenum, 213–45.

Welzer, H. and Wiegandt, K. (2014). *Wege aus der Wachstumsgesellchaft*. Frankfurt am Main: Fischer.

Werner, J. and Dorfner, T. (2013). *Making the Energiewende a Success Story Thanks to Strong Local Authorities*. Part of the Lord Mayors "Sustainable City" dialogue, Text No. 43, May 2013. Rat für Nachhaltige Entwicklung. Available at http://www.nachhaltigkeitsrat.de/uploads/media/Brochure_Making_the_Energiewende_a_success_story_thanks_to_strong_loc_auth_text_No_43_May_2013.pdf.

Wheeler, S.M. (2003). The evolution of urban form in Portland and Toronto: Implications for sustainability planning. *Local Environment*, *8*(3), 317–36. doi: 10.1080/1354983032000098159.

Wheeler, S.M. (2004). *Planning for Sustainability: Creating Livable, Equitable, and Ecological Communities*. London: Routledge.

Willers, B. (1994). Sustainable development: A new world deception. *Conservation Biology*, 8(4), 1146–8.

Wittek, S., Feindt, P.H., Gessenharter, W., Hoppe, J., and Seifert, E.K. (eds). (2002). *Nachhaltigkeitsindikatoren und Partizipation*. Hamburg University Press. Retrieved October 23, 2010, from http://hup.sub.uni-hamburg.de/opus/volltexte/2008/45/pdf/HamburgUP_Wittek_Nachhaltigkeit.pdf.

World Commission on Environment and Development [WCED]. (1987). *Our Common Future*. Oxford: Oxford University Press.

Yin, R.K. (2009). *Case Study Research: Design and Methods*. 4th edition. Thousand Oaks, CA: Sage Publications, Inc. Kindle edition.

Yli-Viikari, A. (2009). Confusing messages of sustainability indicators. *Local Environment*, 14(10), 891–903.

Young, M. (2001, 29 Jun). Down with meritocracy. *The Guardian*. Retrieved March 31, 2015, from http://www.theguardian.com/politics/2001/jun/29/comment.

Zahrnt, A. (2014). Teatime with Angelika Zahrnt. *Open Citizenship* 5(1), 122–7.

Zhang, Y. (2013). Urban metabolism: A review of research methodologies. *Environmental Pollution*, *178*, 463–73. doi:10.1016/j.envpol.2013.03.052.

Zimmermann, M. (2007). Local governments and sustainable development. *Environmental Policy and Law*, 37(6), 504–7.

Zwangsräumung verhindern. (2014). Bündnis Zwangsräumung verhindern – Platform Impeding Evictions, Berlin. *Open Citizenship*, 4(2), 122–5.

Index

Note: Case study descriptions in the six cities are written in **bold**.

For Product Safety Concerns and Information please contact our EU representative GPSR@taylorandfrancis.com Taylor & Francis Verlag GmbH, Kaufingerstraße 24, 80331 München, Germany

Batch number: 10725973

Printed by Printforce, the Netherlands